P9-DVS-148

Darwin's Dreampond

Darwin's Dreampond
Drama in Lake Victoria

Tijs Goldschmidt

Translated by Sherry Marx-Macdonald

The MIT Press
Cambridge, Massachusetts
London, England

Originally published in Dutch as *Darwin's hofvijver: Een drama in het Victoriameer* (Amsterdam, The Netherlands: Prometheus, 1994). © 1994 Tijs Goldschmidt

Publication has been made possible with financial support from the Foundation for the Production and Translation of Dutch Literature.

This book was set in Sabon by The MIT Press.
Printed and bound in the United States of America.

Library of Congress Cataloging-in-Publication Data

Goldschmidt, Tijs.
[Darwins hofvijver. English]
Darwin's dreampond : drama in Lake Victoria / Tijs Goldschmidt : translated by Sherry Marx-Macdonald.
p. cm.
Includes bibliographical references and index.
ISBN 0-262-07178-9 (hb : alk. paper)
1. Lake ecology—Victoria, Lake. 2. Ciclids—Victoria, Lake—Evolution. 3. Victoria, Lake. I. Title.
QH195.V5G6413 1996
575.01'62—dc20 96-24478
CIP

Contents

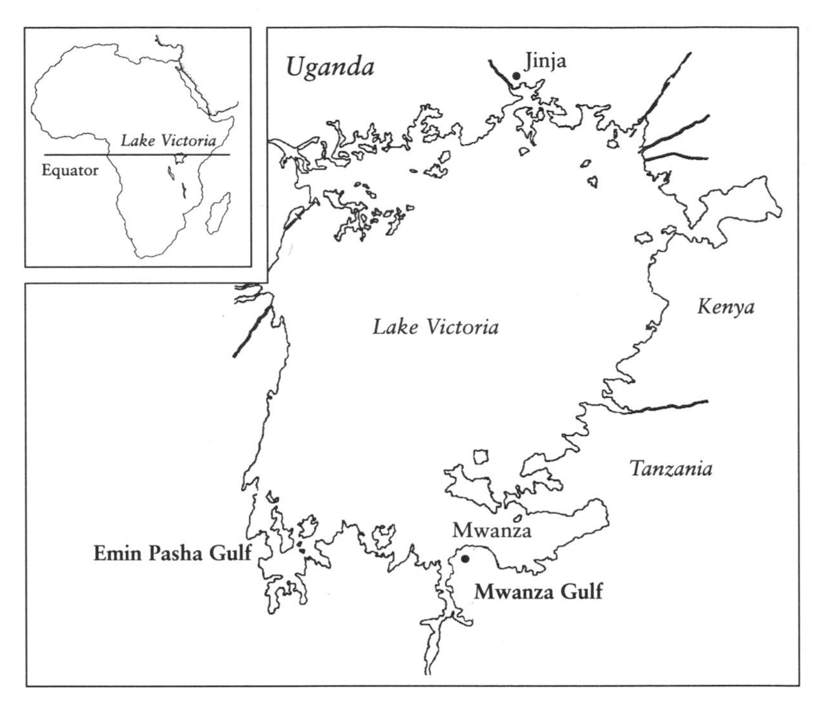

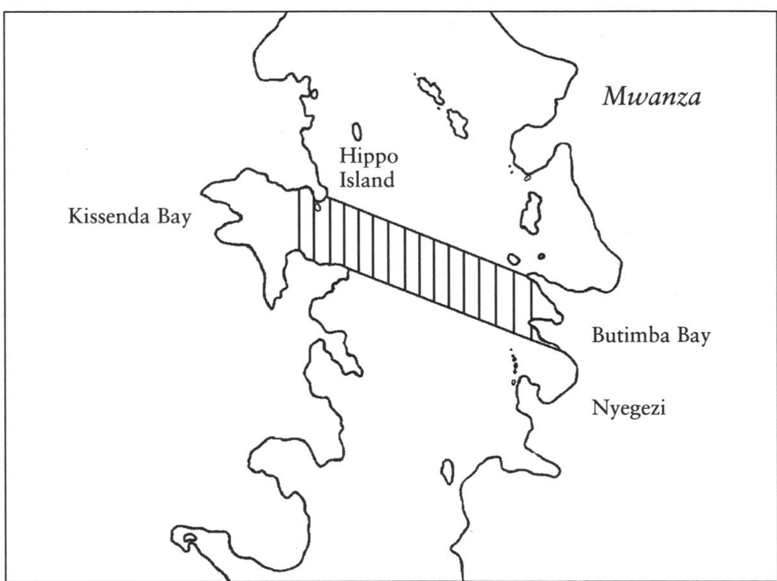

Figure 1.1
Lake Victoria, with the transect of Mwanza Gulf from which samples were taken over a period of years.

1

The Shock

The net disappeared slowly into the water. Three red plastic balls, the floats, bobbed up to the surface about forty meters behind the wooden boat. Mhoja accelerated a little until the iron cables joining the net to the boat pulled taut. When they emerged, dripping, from the water, he looked silently in my direction. I drew a circle in the air. Mhoja looked at his watch and turned the engine up full.

It was the rainy season, 1985. We were journeying up Mwanza Gulf, a southerly offshoot of Lake Victoria. The dragnet would be pulled in after ten minutes. Elimo, who had been bailing out the boat with a rusty Africafé can, sat down next to Mhoja. They stared through the floorboards at the bottom of the boat. In a short time, it had changed into a labyrinth of tunnels, the collective eating efforts of thousands of wood-boring insect larvae. Mhoja pulled out a piece of raw cotton from his trousers pocket and stuffed it into a hole with a stick. Meanwhile, I contemplated what would be worse—to continue the journey in this sieve of a boat, bailing all the way, or to stop work and have the boat repaired. A repair that would doubtless take a long time. Gasoline had been rationed for years. Now that the country finally had some fuel, the boat I was in leaked.

Not wanting to spoil the day, I cast my gaze upward. A seventeenth-century Dutch sky hovered above Mwanza Gulf.

"Look," I said. "*Mawingu kama picha ya mbwana Salomoni Ruysdael, m'Holanzi*, a sky like one in a painting by the Dutchman Mr. Salomon van Ruysdael."

"*Sawa, sawa*, not bad," said Mhoja and Elimo affably. Without pursuing my remark, they continued their guttural conversation in Sukuma. It was amusing. Now that I could effortlessly follow their conversations in Swahili, they reverted to their native language, which I could barely understand. Should I learn Sukuma or politely abandon the pursuit? The language interested me. I hadn't come across a Sukuma grammar book anywhere, but Clementi, a French-Canadian priest who had spent half his life here, had promised me a copy of his stenciled notes. He had pointed out an interesting detail: when the men counted in Sukuma, they used different words than did the women, and young cattle herders used different words again.

Elimo and Mhoja finished their throaty exchange about the price per kilo of brown beans. I let my arms trail limply in the water and looked down at my hands. For a few moments the only sound was that of the outboard motor, until it was interrupted by the call of an African fish eagle high above. Elimo and Mhoja resumed talking, this time broaching their favorite subject: the price of a woman. It was a shame I couldn't understand them better. Elimo had his sights set on a beautiful, serious girl called Maisha—a walking fertility symbol, and priced accordingly. They took turns clicking their tongues on their palates, then shook their heads, repeatedly uttering short, clipped "uhs." How could Elimo acquire enough cattle to satisfy her parents? He didn't want to spend the rest of his life working as a serf on his parents-in-law's land to pay off the dowry. If only his family could help. And why were her parents so difficult? It was typical of people who lived far from the city. Dowries were no longer common, but Elimo had resigned himself to having to pay and was saving as fast as he could. You could see it in his greedy glance every time the net surfaced: here, that big fish, give it to me ... an eye, a little finger, the voice ... It was as if he were conjuring up the coveted Maisha as he fished. He sold his catch and, as soon as he had enough money, converted it into Sukuma currency–cattle. While bank notes just lay around in piles gathering dust, cattle bore young, the Sukuma reasoned.

As long as Elimo and Mhoja could do what they liked with the marketable fish, they had no objections to my using the small, very bony specimens for my—they were too courteous to say it—crazy work. They never could understand how I could be so enthusiastic about catching nothing after a whole day's fishing. That the absence of fish could be as important to me as their presence was less comprehensible to them than Ruysdael's *Bleaching Grounds near Haarlem*.

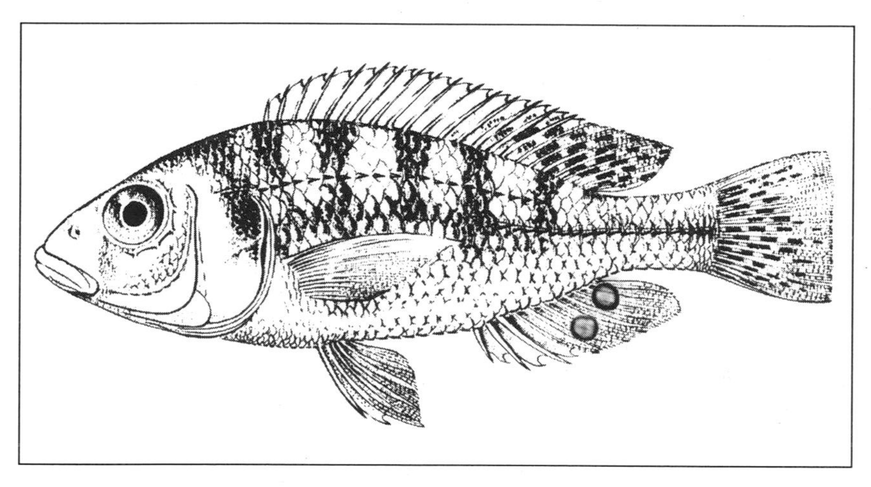

Figure 1.2
A typical furu.

Lake Victoria is a freshwater sea. A shallow saucer filled with water, about the size of Switzerland. It is transected by the Tanzanian, Kenyan, and Ugandan borders and scalped in the north by the Equator. Mwanza Gulf is not very large, about the size of the Thames estuary. During the early 1970s, the Dutch government became involved in efforts to develop the fishing industry in this region. In terms of number of species and biomass, the fish fauna of the lake was known to be dominated by a group of small perchlike fish, the cichlids of the species-rich genus *Haplochromis*. It was on these fish that attention was to be focused. Technicians were dispatched from the Netherlands, together with a boat, a secondhand shrimp trawler. The boat was transported from the Netherlands to Mombasa by ship, and from there by train. Strapped onto an open railway carriage, it traveled straight across Kenya to Lake Victoria. To the Masai, who had never seen such a seaworthy ship before, it was the umpteenth Western idea to rumble across the savanna. The ship sailed under its own steam from Kisumu to Mwanza, where trawler fishing had begun in 1973 under the supervision of a Dutch fisherman. On board the trawler, Tanzanian pupils from the fisheries school at Nyegezi, a small village near Mwanza, became acquainted with Western fishing techniques.

During the early years, the teaching vessel *Mdiria* (Kingfisher) was the only trawler to plough the waters of Mwanza Gulf, but the Tanzanian and Dutch governments had plans for a major expansion of the fishing industry in this area. Trawlers were built with Dutch funds earmarked for foreign aid and a fish-meal factory was to be built that would process sixty tons of cichlids per day into animal forage. Sixty tons of cichlids per day? How many days would it take before the fish supply in Mwanza Gulf was depleted? Had anyone taken that into consideration? Had the size of the cichlid population been assessed? Was anything known about the resilience of the ecosystem? How many tons of cichlids could be caught each year without permanently damaging the ecosystem? As it turned out, very little was known about any of these things.

News of the plan reached the ears of two zoologists from the University of Leiden in the Netherlands, Gerrit Anker and Kees Barel. They were shocked. To what extent would the vulnerable fauna of Mwanza Gulf be able to withstand such heavy fishing? These reclusive scholars felt compelled to oppose the plan and so traveled, quite against their nature, to East Africa. They had become as attached to their laboratory in Leiden as barnacles to a rock and really only wanted one thing: to return home as quickly as possible. But they realized this too late, and thus found themselves journeying across the paradisaical savanna, arguing all the way, in the footsteps of the nineteenth-century "discoverer" of Lake Victoria, Speke, and his companion, Burton. Barel longed to start a project in an exotic location, although nothing on earth would induce him to go there; he saw this as his golden opportunity. Anker, the epitome of rigor and reliability but apparently too cautious ever to take a risk, focused mainly on the drawbacks: the blistering heat, the unfathomable Africans, malaria, and the tsetse fly. Sulking near a spring on the Nile—or so the story went back at the Leiden Zoological Laboratory—these two scholars decided that the Netherlands was indeed a wonderful country. They returned home. What to do now? Barel came up with the perfect solution. In 1977, staying behind in Leiden himself, he stationed at Mwanza a group of young colleagues with a passion for fieldwork. These researchers and their successors, together with Tanzanian biologists, studied the ecosystem of Mwanza Gulf and the surrounding area. In the meantime, the anatomists at Leiden continued to study these fish on the home front, receiving specimens of any species they wanted from the field. This was, in short, the background for my own "bobbing around" on Lake Victoria from 1981 to 1986.

Very little was known about the biology of Lake Victoria's cichlids in the mid-1970s. Humphry Greenwood, former curator of the Museum of Natural History in London, had made describing these fish his life's work,[1] but the specimens he had collected during the 1950s had come mostly from the Ugandan side of the lake. We ran aground trying to identify the species on the Tanzanian side. Greenwood's key for identifying species did not work for Mwanza Gulf and surrounding areas. His descriptions were inadequate, although it was hardly his fault. Over the years, an endless stream of nameless species had surfaced. The taxonomists of the Haplochromis Ecology Survey Team (HEST), as Barel's group called itself, identified a total of one hundred and fifty new species, enthusiastically during the first years, almost reluctantly later on, as nothing is more stultifying than discovering something unique every week. I myself once caught an unnamed, exceedingly lively, fiery-tempered male with purple flanks and a pitch-black mask but then let him go because at the time I couldn't face discovering another new species. I don't think it was ever caught again!

It was difficult to observe the behavior of the haplochromine cichlids in the turbid 25-degree broth of Mwanza Gulf. On the odd occasion, when there wasn't a breath of wind, I did manage to watch them digging a nest pit, blushing purple or orange, embroiled in fierce combat, or engaged in consuming a worm. It was these moments that kept me going. The underwater biotic community did indeed exist, but who were the cichlids? The various species resembled each other closely and their number was dauntingly large. It took months to become familiar with one particular group. Classification, the starting point of all ecology, was not easy. Some never mastered it, but even those with an eye for the shape, structure, and color of these fish had to exert themselves and regularly practice their skills. After a holiday of several weeks, it would take several days before a taxonomist felt at home again with his specific group of cichlids.

In several places in his book *La pensée sauvage*, Lévi-Strauss refers to the considerable taxonomic insight of several non-Western populations:

> [The] extreme familiarity [shown by native peoples] with their biological environment, the passionate attention which they pay to it and their precise knowledge of it has struck inquirers as an indication of attitudes and preoccupations which distinguish the natives from their white visitors.[2]

The last thing I wanted was to be associated with the white visitor of Lévi-Strauss's book. Excited and eager to benefit from the Nyegezi inhabitants' knowledge of the cichlids, I showed them the most diverse species of *Haplochromis*: large and small, elongated and compact, full- and thin-lipped, wine-red, canary-yellow, pitch-black, but every time I received the same reply: "This fish, the name of this fish? It's *furu*, it's called furu."

"And what else?"

A long silence usually ensued, but then it came out again, solemnly and politely: "*Mzungu,** wanderer, you've caught a furu. *Haki a Mungu*, I swear it."

The villagers of Nyegezi had names for various fish, such as the tilapia, the elephant snoutfish, the walking catfish, and the lungfish, but their knowledge of the many different species of cichlids was limited. The Sukuma were traditionally farmers and herders, not fishermen. Perhaps this was the explanation. Except for Greenwood, African and Western biologists had been all too happy to leave the furu (primarily of the genus *Haplochromis*) for what it was: an impenetrable and difficult group.

What, then, was our reason for wanting to study these creatures? In the first place, it seemed useless to draw up a fishing policy without first understanding the ecosystem. But there was also a selfish motive. The furu of Lake Victoria fascinated anatomists, ecologists, ethologists, and evolutionary biologists alike.[3] We believed that Lake Victoria might be harboring a species flock: a group of closely related species that had descended, during a relatively recent past, from a common ancestor.[4] A species flock is ideal for anyone interested in comparative research. Comparisons of species, whether pertaining to shape, physiology or behavior, are not confounded by fundamental differences in the body plan of the organisms. Moreover, we suspected that this species flock had not yet crystallized into its "final" form but was dynamic. New species were probably still joining its ranks every day.

In themselves, these were not decisive reasons for journeying to Lake Victoria. We could have gone instead to the Galápagos Islands. The prototype of a species flock inhabits these regions: Darwin's finches, distrib-

*The original meaning of this word was apparently "wanderer." It was the name given to the first Westerners who traveled through East Africa. The word later came to mean "European" or "Westerner."

uted amongst the different islands. These finches descended from a small group of finches that had swept over from the South American continent and settled on this remote group of islands, where they had discovered an uninhabited terrain with virtually no rivals. At most, they got in each other's way.

These initial circumstances, combined with the geographical isolation of the islands, led to the emergence of thirteen species of finch. A fourteenth species also appeared on the more northerly Cocos Islands. Darwin's finches differ greatly from each other, particularly in the shape and structure of their beaks and in their manner of hunting for and digesting food. They have radiated, fanned out into different specializations. The official name for this phenomenon is adaptive radiation. They were certainly a suitable group for comparative study, but they had already been singled out by many biologists.[5] Another isolated group of islands with an avian radiation—some of whose species have since died out—is Hawaii. Twenty-two living species of honeycreepers (Drepanididae) as well as several that recently became extinct developed there from a common ancestor, each adapted to hunting for and digesting different kinds of food. They, along with many others that could also be cited, represented other candidates for comparative research. But not one of these groups had radiations as exceptional as those of the cichlids.

In the East African rift—a long, narrow fault-induced valley in the earth's crust—lies a chain of deep lakes: Malawi, Tanganyika, Kivu, Edward, and Albert.[6] Each is an evolutionary laboratory in its own right, with, like Lake Victoria, its own radiations of cichlids.[7] The radiations of Darwin's finches and honeycreepers utterly pale in the face of cichlid radiations. These are the most spectacular in the world, giving the impression of having been created by a masterly magician who now and then feels the urge to outdo himself again. One or, at most, several cichlid species go into the hat and every time, without fail, dozens or even hundreds come out—displaying the most unlikely shapes, colors, and behavior. The trick has been most successful using cichlids, but could, if necessary, be performed using other species, such as catfish, snails, ostracods or sponges, all with smaller radiations.[8]

The recipe is quite simple: allow a saucer to form in the earth's surface, or gently tear open the surface (not too deeply, as excessive warmth from the earth's mantle can spoil everything). Let the saucer or fissure fill up with river water. Ensure that several riverine cichlids, preferably of the omnivorous kind, find their way into the virgin lake, and that's it: the rest will take care of itself. After, at most, several hundreds of thousands of years, anywhere between dozens and hundreds of kinds of colorful cichlid species will have developed, closely resembling each other in their basic features but outfitted with anatomical equipment and behavior to allow them to process totally different kinds of food. Several millions of years later the radiation will have progressed even further: the diversity of shapes, colors, and behavior patterns will have become so great that any amateur could easily learn to distinguish one kind from another. The differences will have become as distinct as those between a cow and a pig. Ten million years hence, the differences will have become even more pronounced, and many intermediate forms will have died out in the meantime.

Curiosity about the way this recipe works was one of our main motives for wanting to study the species flocks of the East African lakes. New species were—literally—appearing, changing, and disappearing before our very eyes.

The East African lakes differ in age, and several, as far as we know, have never been linked with other lakes. This provides a unique opportunity for comparison: cichlid radiations can be observed at different "stages" of evolution, in this case, as living organisms. The work of comparing the different evolutionary stages of a particular group is normally the domain of paleontologists. But they have to make do with a skull, bones, and scales, and reconstruct the rest. Behavior, telling colors, soft structures, and physiological features simply do not fossilize.

While the recipe for determining the origins of a radiation is perhaps simple, establishing how it actually works is quite another matter. Dozens of evolutionary biologists have busied themselves with this question throughout our century, focusing initially on classifying and describing species and charting their patterns of distribution. But this is where the problem starts. How can they be distinguished from each other? Which organism appears where, and when? Some biologists have attempted to establish whether the furu are entitled to be called a species flock: have all the furu within the confines of the Victoria basin really descended from a single common ancestor? And if so, how? In accordance with which mechanisms? Has natural selection played an important role, and if so, has it been responsible for the appearance of different forms, new species, or both? What role has sexual selection—that distinct form of natural selection—played in the emergence of new species? And what do the hundreds of furu species actually do; what is their role in the ecosystem? How do you explain the fact that so many species exploit the same food source without competition for space or food necessarily becoming fatal? It is strange that it has always been the furu that have developed new species and radiated at an explosive rate. Why hasn't the "little sardine," *Rastrineobola argentea,* ever done this? Or the lungfish, for that matter? Isn't it about time for it to renew itself? It has scarcely changed in over a million years.

I thought there was something unique about these beautiful little fish that fit nicely into the palm of your hand. I even supposed that the radiations of these organisms in the East African lakes might serve as a model for many older groups of fish, as well as for birds and mammals.[9] A young, radiative group in an isolated microworld. The key, perhaps, to a better understanding of evolution.

While I was preoccupied with these questions, whose answers were partial at best, Mhoja and Elimo were busy—in the same boat fueled by the same gasoline—"saving up" for one woman after another. Before long they would have a farm, fields, four wives, and sixteen children, and would organize dance competitions, paint their bodies, and perhaps even carve life-size expressionistic statues. How you used your gasoline—that's what it all boiled down to.

Total Destruction?

Silently, Elimo went and stood before one of the hand-winches. I walked over to the other one. Mhoja, forever at the helm, looked at his watch again. In thirty seconds, the ten minutes would be over. As soon as Mhoja gave the signal, Elimo and I started turning—by hand—the cables of the creaking winches. After years in Mwanza Gulf, I was so familiar with the environment that it wasn't difficult to predict which furu would be caught where. This time, too, I was expecting certain types and had decided how many samples of each would go to the institute. Elimo and Mjoha would take care of the rest of the fish.

Elimo's face clouded over as soon as the net came aboard. The catch was abnormally small. Disgruntled, he untied the end of the net and shook its contents into a plastic bucket. The catch consisted mainly of furu, but the longer I looked into the bucket, the less I understood about where we were. Something wasn't right. Was I familiar with these species? As if drunk, I looked around, trying to orient myself. Directly in front of us were Mwanza Bay and Butimba Bay. Behind us was Kissenda Bay, bordering on the papyrus swamps. This was indeed Station G, the location in Mwanza Gulf from which most of the samples had been taken, but the moment I looked into the bucket, I felt as if I was somewhere else. More northerly? I recognized the species, but they weren't from here. And species that had always been caught at this location were absent—for the first time in all these years. Might the recent invasion of the Nile perch have something to do with this? For more than a year, fishermen had been catching this introduced predatory fish in large quantities. At first they thought a monster had invaded Mwanza Gulf. The gill nets were full of enormous holes: the monster plowed right through their delicate, fine-meshed nets.

Nile perch live mostly off furu and grow to enormous sizes. Some specimens weigh more than seventy kilograms. Had the fish I had been studying for several years suddenly disappeared, following an earlier slow decline in their numbers? It certainly appeared as if a limited number of species of furu from deeper waters had taken over their habitat. Laughingly, Mhoja said: "Don't worry, your fish have just gone away for a while. You go away for a while sometimes too, don't you? I'm sure they'll be back." I wasn't so certain. Had the organisms we had named and in some cases even become acquainted with disappeared into the jaws of the Nile perch? Perhaps after an initially slow start, the Nile perch population was suddenly increasing at an explosive rate.

Above water, everything appeared deceptively normal. Several gray kingfishers with patches of black-and-white plumage fished in the boat's wake, while the little ferryboat owned by the German gold seeker chugged across the horizon. Since Tanzania had become independent, the gold seeker hadn't been allowed to hunt for gold, only to sail back and forth. Nearby, a lungfish stuck his prehistoric green and black head out of the water, filled his lungs with air, and vanished into the depths to continue his murky existence. But how long would this last? Was everything, including the furu and other indigenous species, being destroyed by hordes of giant Nile perch? If so, all it had taken to cause the total disruption of the ecosystem of the largest tropical lake in the world was a man with a bucket.

It was unfortunate for these fish that they weren't warm-blooded, that they were hairless, and that they inhabited turbid waters. Little attention was paid to them. But I couldn't reconcile myself to the fact that they might disappear unnoticed. Nobody could have predicted in detail the changes taking place in Lake Victoria. In retrospect, some of the changes can be explained. But let me begin by describing what I saw there, what it was that made these fish so exceptional.

2

Everything but an Eye-Biter: A Branching Out of Forms

Mwanza, 1981

First, the journey. It seemed as if it would never end. With Aeroflot to Moscow, to Aden via Cairo. Irritably hanging around airports and only days later reaching the final destination, Dar-es-Salaam. A four-day wait, then on to Mwanza in a phantom plane, a virtually empty aircraft that I had repeatedly been told would be overflowing with passengers; every seat was reserved. Only after insisting for days was I able to book a seat. Where were all the people with tickets? Were we already airborne? I heard nothing and felt even less. My legs had fallen into a deep sleep and the numbness was slowly but surely creeping up toward my waist. Why this voluntary exile? Was I up to it—arriving somewhere new, claiming a place for myself, creating my own niche?

The plane started flying lower and, after a short time, lower still, as if slowly descending a staircase carpeted with fog. Ahead of us—the blue lake, papyrus swamps lining the shores, and marshy fields. Palm trees scattered like ink flecks along red roads and pathways. Mud huts and small stone houses with corrugated iron roofs. Africans growing more life-sized by the minute, filling out. And cattle egrets, absolutely motionless, as white as if they had been perforated out of the green backdrop. Was I being seduced by a landscape?

The plane repeatedly touched the airstrip, haltingly putting an end to the dream. But as soon as it came to a standstill, the dream started all over again: a caressing wind at the top of the stairs. Heavy, oily odors, and an agreeable warmth. If only I could sit here for a moment to contemplate Mwanza before merging into it. I felt a nudge in my back.

I descended the stairs and strolled in the direction of the arrivals hall. What followed was a disconcerting encounter. Who was standing there waiting to greet me? The very people I thought I'd escaped by leaving the Netherlands. They were smiling at me, and the longer they did so, the more convinced I became that one day they would never do so again. The Europeans, mostly sunburned compatriots, closed in around me. One of them even touched me.

"We'll talk about it later."

"They'll never learn."

Voices were ringing on all sides: "Yes, they will, but slowly ..."

"The Tanzanians?"

"Never. We've been trying for fifty years ..."

We, they, it? Where had I landed? Were there any Tanzanians around? All I saw were the disenchanted Europeans who encircled me. Ah yes, thank goodness: a lone black head moved like an electron around this white nucleus of moaners, giving us a wide berth. Had I traveled eight thousand kilometers to end up caught in a net of Europeans? I escaped to the washroom. Come now, five years in a lifetime is nothing, I said to myself, don't always act as if you've just arrived in Bergen-Belsen. There are bound to be some nice people among them. But do make clear right away that you don't like games.

Violently retching into a toilet overflowing with feces but devoid of water, I sought consolation in the thought of the silent head I'd spotted among those of the Europeans: wild, dark brown curls, a roundish face with alert eyes. A fiery temperament, like Rembrandt, a young Rembrandt who had ended up in the health-care system or fertilizer industry.

Within half an hour, I was given to understand that I had been identified as "single" and that it was better not to be so. In reply to my inquiry about accommodation, it was made very clear that I didn't have a hope. Singles were only eligible for one of the houses belonging to the Dutch community after all the families had been royally accommodated. Until then, they were obliged to stay in a hotel at their own expense, or arrange housing for themselves. I didn't mind too much. I had no great desire to live in a fashionable enclave among Europeans.

By chance I heard that a retreat owned by the Order of White Fathers was vacant. Its roof leaked but something could be done about that. It was situated on the highest point for miles around, at the top of Luguru Hill, about twelve kilometers from Mwanza. No expatriate wanted to live on the hill because once, many years before, a clergyman had been killed there by thieves. The top of the hill was known as an isolated, dangerous place. I was keen on renting the caretaker's house, but a missionary at the diocesan office told me the mission wanted to keep it for a family. Upon hearing this, I walked into town. I was afraid I might end up in a depressing hotel room in the dusty center of Mwanza. But strolling through the streets, I conceived a plan that I implemented a few days later. I borrowed a serious-looking woman and her small child from a male expatriate. Flanked by the mother and holding the child's hand, I again asked "if it might not be possible ..." A friendly Dutch missionary—not the one I had encountered on my first visit—and a robust-looking nun took pity and the stray young family finally had a place to stay.

It was now 1982 and I had been living on Luguru Hill for several months. As it turned out, I wasn't alone. Levocatus Enoka, the guard, watched over the buildings of the retreat for the mission. Moreover, he was a *mganga,* or traditional healer, and treated his patients in a building close to the caretaker's house. Mhoja and Elimo also lived on the hill, in a small house a stone's throw from mine. Once in a while, guests would appear, among them scientists from European or American universities: biologists, paleontologists, and anthropologists. These visitors often felt lost—particularly if they were in Africa for the first time—and would cling to me immediately after their arrival as if rescued from drowning. It was a good thing, because I was longing for kindred spirits and in this way there were always some around. Because they were better informed of the latest developments in our field than I was, they became an important source of inspiration for me. Besides, they brought cheese, liquor, and batteries with them.

It was late afternoon. After work, I walked home from the institute with Isaac, an anthropologist from Leiden. We passed some fields belonging to small farmers. There had scarcely been any rain for the past month and the maize was a sorry sight.

"Why don't the farmers irrigate their fields?" asked Isaac. "It's only a hundred meters to the lake."

"They don't have the money for a pump system, or the pump gets stolen or breaks down and there are no parts. Or thieves carve bracelets out of the plastic pipes that carry the water. It's risky to undertake anything," I replied.

"But why don't they form a human chain and pass buckets of water up from the lake? It would only take an hour a day and their worries about crop failure would be over."

We ran into a good-natured woman farmer in her field. We chatted a bit and then I raised the subject.

"It might still rain," she said optimistically. "Perhaps even this evening."

We burst out laughing: "Yes, you never know."

The woman's reply struck us as fatalistic. Was there such a thing as an Islamic attitude?

> Allah possesses all human qualities in the absolute sense: He knows more than the most learned scholar. He is the Judge of all Judges. He is the best Protector, the indisputable Savior; if farmers beg for rain, they address their plea to the Noble One, the Giver, the Bestower, He who makes grain grow, He who creates and destroys all things. This brings us to a fundamental Islamic belief, one that is often called fatalism: *Mango Mungu atendaye,* or, as rendered in a Swahili proverb: "God is the maker of all things." This should be taken literally: there is nothing that man does himself without the Lord first having willed it. What He wills, happens; what He does not will, does not happen. Man has no initiative, he acts only on the will of God. ... Thus, each man is Abd-Allah, a slave of God, and man's primary activity—far more important than that of plowing or sowing—is to praise the Lord through the endless reiteration of his many glorious names, those numerous renderings of God's infinite greatness.[10]

We took leave of the woman and started up the hill.

"Much depends on rain. The people speak of it as if it were a person: how the Rain descends upon them and falls, much welcomed, or lies low behind the ridge of hills. Did I tell you about the rain gauge?" I asked Isaac. "For months, early in the morning, I had been measuring how many inches of rain had fallen. I would note the reading, empty the container, and then return it to its holder. One morning, after a heavy downpour, I eagerly went to take the reading. To my great surprise, the container was empty."

Isaac looked at me inquisitively.

"Mother Theresa, who occasionally helped in the housekeeping, thought the glass was meant to be emptied and had been trying to save me work. When she had arrived in the morning, the first thing she had done, before I awoke, was empty the container. The problem was that it was impossible to find out how long she'd been doing it. And so science progresses. Slowly but befittingly."

"What were you actually looking for?" asked Isaac.

I explained: how many inches of rain fell, how strong the wind was, how deeply the light penetrated into the water, and its wavelength. In this way I hoped to be able to determine why most furu brood only in certain seasons and not throughout the whole year. In the dry season, when the winds were strong, the water in the lake mixed completely. The stratification disappeared. The water temperature and oxygen content were virtually the same from top to bottom. The wind churned up the muddy lake floor, so that the nutrient salts that had sunk to the bottom were absorbed into the water again where they could be eaten by algae. The algae proliferated as a result: from one week to the next, the water suddenly turned green. The growth of algae signified an abundance of food for the zooplankton that lived off them. The zooplankton subsequently flourished. Most furu brooded just before this blooming of plankton. The female fish that brood their eggs in their mouths released them at precisely the right moment—so there was enough food for the fry to eat when they left their mothers for good. Someone had come up with a plan to ask the Tanzanians not to fish with commercial trawlers during this brooding period, to give the new generation of fish a chance to survive.[11] But more details about how the ecosystem functioned were needed first. It would be dangerous to establish a fixed date as the beginning of the brooding season because each year was different—there was no such thing as a normal year. I often spoke with local biologists of my intention: to assess—on the basis of the number of rotations made by a small windmill in an hour—how long it would take before all the female furu would be swimming around with their mouths full of eggs. I think they liked the idea. There was something magical about it.

"Do the Tanzanians have any affinity at all with what you're doing?"

"Not much," I replied. "Most people, even the well educated, are too busy trying to survive. As long as a person is still struggling to procure two meals a day, there's no time to contemplate the theory of evolution. Fish is, of course, a natural source of protein. The Tanzanians are interested in the practical applications of our work, but no one has any time for the theory of evolution or nature conservation."

I picked up a crumpled piece of paper from the sandy path and read aloud: "How sugar is produced from sugar cane. Sucrose or sugar is manufactured in the leaves of the plant by photosynthesis ..."

A few moments later Isaac bent over, picked up a scrap of newsprint, and read: "Our objective was to cultivate a further and deeper understanding of Socialism and Self-reliance. The Party ideology and its implications ..."

The sandy path was clean. But along the edges lay scraps of paper, fruit rinds, and the odd piece of plastic. I chimed in as soon as Isaac had finished, reading aloud from another piece I had just retrieved: "*durch der Engel Hallelujah, tont es laut von fern und nah: Christ der Retter ist da.*"

Suddenly we were surrounded by a group of small children eagerly helping to look for scraps of paper.

"Here, *Mzungu*, wanderer, for you," said a shy little boy laughingly, dressed in nothing but a hole-ridden T-shirt. He handed me a piece of cardboard.

I thanked him and read aloud while he jumped up and down with joy: "It is fantastic what money can buy: TROPICANA."

The letters were printed above a photograph depicting two cigarette-smoking Africans clad in black suits. They were wearing white leather shoes and seated on a settee in a garish interior. Searching, collecting, and taking turns reading, we walked up the hill along the sandy path. We didn't find a single piece of text in Swahili or Sukuma. The procession of children accompanying us steadily increased.

"This is the only real problem," I said to Isaac. "If there were eight million Tanzanians instead of the present twenty million, and this number didn't increase, everything would be all right. But the population is growing annually by more than 4 percent. Malthus[12] would have turned in his grave if he'd seen this horde of children emerge from the woods. The hell he predicted can't be far off."

The children couldn't get enough of greeting us. "Good morning, father" and "Good morning, teacher" echoed on all sides.

Upon reaching the top of the hill, we walked past the chapel ruins, of which only remnants of the occasional wall were still standing. Kiokikote, a friendly little stray dog who had showed up one day and been with us ever since, ran up excitedly to greet us. I kicked at a sawed-off gas cylinder that had once served as a church bell, adding: "I wanted to build a chicken coop here but the Tanzanians vigorously opposed the idea. No chickens on sacred ground. I abandoned the plan."

On the reddish-brown tile floor of the former house of God lay a pair of rusty scales, surrounded by pieces of iron pipe and decaying reed. Next to the scales, an iron cross protruded at an odd angle out of the rubbish. A thick line in white chalk had been drawn on the tile floor depicting a large rectangle, but just before it collided with its starting point, it retreated and dived inward, into the seclusion of the rectangle. After following an uncertain zigzag path, it came to a dead end in the crook of a V, encompassed by five circles, two small ones and three larger ones. Lizards with purple heads and dark blue bodies scuttled back and forth across the drawing, making jerky vertical movements with their heads and scooting away as soon as the specter of man loomed up on their prehistoric retinas.

This place, which seldom received visitors, was teeming with animals. Yellow-spotted dassies, distant members of the elephant family resembling large marmots, sent off alarm signals when we approached too closely. They ran straight up the rock face and, from a safe height, sat there motionless, like a set of gray tea cozies. Isaac couldn't see them. I pointed them out but he still couldn't see them. Then he suddenly spotted them. Three, seven, dozens. Now he couldn't *not* see them. There was a "search image" to focus on. Noticeably absent was the leopard. Previously, when there had been a substantial dassie population, the leopards had kept their number down. But since the leopards' disappearance, their numbers had increased explosively. I made Homeric comparisons using the following ingredients: overpopulation, leopard, man, dassie, Catholic church.

The clear blue sky was flecked with spots. Black kites glided high overhead, tracing a windy route of long continuous lines within the confines of the warm air current on which they floated. Lyrical, perfectly articulated lines.

A panoramic view of the surroundings could be had from this highest vantage point. To the east, rugged, rocky Sukuma country; to the south, Mwanza Gulf, which only branched off after lengthy indecision. Looking back, the indecision was justifiable. Which African body of water would want to be called "Smith's Arm" or "Stuhlmann's Arm"? Which body of water would want to be named after the organizer of a grueling expedition, even if he had perhaps earned his spurs in zoology by occasionally raking his fishing net through the water and sending a few exotic water fleas back to Bismarck's biological staff? Vistas of the gulf could be seen to the north and west and, in the distance, the lake—a veritable sea. On a clear day, Ukerewe Island was visible as a thin strip along the horizon.

We stood up and walked past my house.

"It reminds me of one of Palladio's small villas," said Isaac.

"Only less comfortable," I replied.

The roof still leaked in several places. But it hadn't bothered me much because, since the April downpour, there'd been very little rain. Only the odd shower. On those occasions you had to make sure all the lights were out. When there was a power failure, I always checked that the cassette deck was on "Play" so I would know when the power returned. "*Nur weiter denn, nur weiter*" would then slowly become audible, although sometimes the current jumped suddenly to more than 300 volts. I didn't know whose baritone voice it was that sounded so warm and versatile at 220 volts. If the voltage rose above 250, the voice became a shrill cry and touched something in Kiokikote. He would begin a most heart-rending wailing, as if images from a hellish past had suddenly returned to torment him. Within a short time, dozens of dogs for miles around would join in. Howling—wolflike—to Schubert.

Occasionally, but not often, the sounds of ecclesiatical singing emerged from the retreat. Usually it was deathly still. Most of the buildings were vacant and slowly but surely being dismantled. First the windows, then the iron roofing, and so on. I felt surrounded by skeletons. The process of decay was furthest advanced in the building housing the priests' sleeping quarters. It was overgrown with plants and mold. The damp walls were heavily infested with lizards. Birds had built nests on the wardrobes. Monkeys ran through the corridors, transforming derelict bed springs into trampolines. The hollow, orangish-red bricks of the outer walls housed black-spotted emerald-green snakes. The bricks had been manufactured in Europe, then brought to Africa by ship and transported from Dar-es-Salaam to Mwanza. A ludicrous journey ... hundreds of thousands of bricks en route to a country rich in river clay. Hundreds of thousands of bricks crossing the sea, floating on the Suez Canal, across the Indian Ocean, toward the Comoro Islands ... Several bricks lost heart and fell overboard, sinking symbolically, waking the coelacanths out of their dreams that had lasted for hundreds of millions of years.

To Isaac, who had since explored the hill, I pointed out the silver-gray water in the distance: the "ladies' lake," separated from Mwanza Gulf by papyrus swamps and a small strip of land.

"Do you see the shimmering surface? The Sukuma warn you not to sail there at night. Droves of corpulent women enter any boat that ventures there. They come aboard. More women than a man could wish to have in his entire life. The boat, crew and all, gets crushed under the weight of female flesh."

"Hey, a raven," said Isaac.

It was Cas, the black raven with the white neck who woke me every morning at seven sharp. He crowed and flew away from the hill toward the gulf, wing tips spread fingerlike.

The Model T

Our study of the species inhabiting Mwanza Gulf began literally from scratch. Other than the research Greenwood had carried out on the Ugandan side of Lake Victoria during the 1950s, there was no other fieldwork to build upon.[13]

The names of most of the species in Mwanza Gulf were nowhere to be found in scientific journals or encyclopedias, any more than information on their role in the ecosystem, their interrelationships, or their origins. There was no reason for us HEST biologists to go to libraries and read volumes of tedious articles. We would write these ourselves later. Instead, we sailed out of Nyegezi Bay in our wooden boat and proceeded to carve open the overflowing lake in all directions. During the odd nocturnal outing we even scooped fish from the water's surface with our hands. The poet Valéry once wrote:

> Nothing is more perfectly constructed or appeals with more charm to our sense of spatial form and our instinct to give shape with our own hands to those forms we would most like to touch, than this ... gem that I am now stroking and whose origins and purpose for the time being I wish to know nothing about.[14]

It was not like this between the cichlids and us. We didn't have time to experience them as they were. On the contrary, as soon as we found a new specimen, we immediately began formulating all kinds of hypotheses to decipher what it was exactly we had in front of us. How could I ever learn to distinguish all these different fish from one another? Were they in fact biological species? Could this not be a new phenomenon, a huge species with hundreds of masks, behind which one and the same genome was hiding? I became increasingly uncertain about what a species was. The longer I reflected on the concept, the vaguer it became. Any grip I had on the system disappeared. Whereas in the beginning I had headed my letters with "*luctor et emergo*" (I struggle and I surface), six months later this had been reduced to "*luctor,*" and eventually, I used no heading at all for months on end.

To avoid drowning in the complexity of the ecosystem, I decided to trace what happened from the moment a fish was caught in one of our nets. A few hours later a label with brief notations floated alongside the body in a jar of formaldehyde: a temporary name, a date, the time it was caught, and an indication of the place. But what had happened in the meantime?

As soon as the catch came aboard, it was emptied into a large, square box full of ice. Later, I sorted the catch—which usually consisted of several hundreds of fish—on long wooden tables in the corridor of the laboratory. It was very difficult not to group them according to a certain color pattern or shape. Instead, they had to be sorted in the eternal rows of two: males and females. First the males were grouped according to similarity of shape and color, then the dull gray females bearing similar color or spot patterns—though more restrained, whispering—were laid beside them.

Classifying these aquatic animals definitely had an arty aspect to it, in which the expertise and intuition of the taxonomist played a role. A taxonomist attempts to assess the relative importance of a large number of properties that cannot easily be expressed in terms of numbers.

In this way he determines the species to which an organism belongs. Unfortunately, many taxonomists denounce this intuitive aspect of their work. Roaming around in the margins of science and often looked down upon by experimental researchers, they long for recognition from the "hard-core" natural scientists. Experimental biologists often forget that without the expertise of experienced taxonomists their hands would be tied. Despite this, everything is being done to render the taxonomist extinct—the taxonomist, the only one who knows the names of endangered species. Before long we will be left only with the statisticians and gene freaks and there will be no biologists around to identify species or observe organisms or plants.

Science or no science, it's neither here nor there to the victim. It irrevocably ends up in a jar of formaldehyde, surrounded by what are presumably similar species. This is how "species in formaldehyde"— taxonomic species—come into being. Taxonomic species are born of the practical necessity to classify organisms. Biologists can do nothing with a disorganized mountain of fish. But this does not mean that the classification they impose on their material reflects a natural order, even though this is their intention. Ideally speaking, the taxonomic species—the artificial one floating around in a dense cloud of formaldehyde, alcohol, or camphor—overlaps perfectly with the population-genetic species—the concrete species comprising swimming, gliding, flying, walking, or totally immobile organisms. If specimens classified as belonging to different taxonomic species do not interbreed in nature or, to put it differently, if there is a barrier to interbreeding between these organisms, then the taxonomic and population-genetic species coincide. But in dynamic ecosystems, in which new species are continually emerging, it is often impossible to identify to which species an organism *in statu nascendi* belongs. So this aspect of the cichlids that intrigued me so much—the emergence of new species before our very eyes—also became a source of frustration.

Specialists working with the concept of species complain from time to time that they no longer know what a species is, yet biological species are often considered the only true taxonomic category. If man were to disappear from the face of the earth, then the human abstractions of kingdom, class, order, family, and genus would disappear with him into the grave, but the biological species would have a good chance of outliving him. After all, species define themselves, and, with the exception of a few parasites and a limited number of species found only in captivity, they are not dependent on man for survival. In fact, they would be much better off without him.

It is a difficult, perhaps even impossible, task to provide an all-encompassing definition of biological species. The description given by sociobiologist and conservationist Edward Wilson for organisms with a sexual reproductive system suffices reasonably well:

> Species are regarded conceptually as a population or series of populations within which free gene flow occurs under natural conditions.
>
> This means that all the normal, physiologically competent individuals at a given time are capable of breeding with all the other individuals of the opposite sex belonging to the same species or at least that they are capable of being linked genetically to them through chains of other breeding individuals. By definition they do not breed freely with members of other species.[15]

In those cases where identifying species of furu presented a problem, we subjected them to a bombardment of measurements that allowed us to make comparisons. The initial comparison was made with a furu from Uganda's Lake George, which had been analyzed down to the smallest bone. Anatomists in Leiden had produced extensive descriptions of the anatomy of this species, considered a sort of jack-of-all-trades, the Model T among the cichlids.[16] They had intentionally chosen an organism with a very common appearance for their descriptions. According to the anatomists, in terms of structure and behavior this *Haplochromis elegans* most closely approached the hypothetical riverine ancestor of the lacustrine fish. Only after detailed basic information had been compiled on the structure of the skeleton, the shape of the individual bones, and the muscles and connective tissue, was it possible to compare the species and establish, with any certainty, the often subtle differences.

An organism can be seen as a three-dimensional jigsaw puzzle. Only when all the pieces fit together and form an integrated whole can it function optimally. One of the characteristics of organisms is the extreme crowding of their internal structures: there is a struggle for space from birth to death. No space is wasted, which means that every change in the shape, size, or position of one anatomical structure has an effect on another structure.[17]

The cichlids of the East African lakes resemble each other closely in terms of design. The same building blocks—bones, muscles, connective tissue, and so on—are present in each of the more than one thousand cichlid species. The spatial distribution of these blocks is also more or less the same. This similarity in design demands an explanation, but so too do the differences. Organisms are not indefinitely mutable; their ability to change is limited. But despite inevitable historical and mechanical constraints, an incredibly large number of furu variations have emerged from a single theme.

The head of a furu is a rattletrap of more than one hundred bones held together by muscles. During the process of radiation, the head in particular has undergone radical changes. The shape of the skull, the teeth, and the mouth differs from species to species and is sometimes so bizarre that it is almost inconceivable that such a head exists. Sometimes I would try to classify a species without examining it thoroughly. By concealing part of the fish, it was possible to identify which external features defined that particular species. It was often the color or spotting patterns. But the most common distinguishing feature was the shape of the head and teeth. It is usually impossible to classify furu without having first examined their heads, both inside and out. The head of the furu is full of architectural tours de force. It was on the head that the anatomists in Leiden focused. There was work for decades to come. It was even taboo to venture into the "lee" behind the head.

Depending on the species, an adult furu varies in length from five to twenty-five centimeters. Like many other fish, it has only one nostril on each side of its head instead of two. The lateral line system, through which fish can detect the transmission of vibrations through water, is interrupted. These features are only visible under the microscope. Immediately noticeable are the yellow spots on the anal fins of the males, which closely resemble the eggs of these fish—an unusual form of imitation or automimicry. Based on the features described above and several other characteristics, the genus *Haplochromis* has been incorporated into the Linnean classification system.

In 1981 Greenwood published a study representing a major revision of the *Haplochromis* genus, which up to that time had included approximately 120 described species.[1] Greenwood divided these species into the old *Haplochromis* genus and several new genera. In total he distinguished nineteen genera in Lake Victoria. Several of them were also believed to be present in other African lakes. Greenwood believed that the species flocks in the Great African lakes formed a super flock. The HEST taxonomists did not accept his revision, which meant that we biologists continued to adhere to the old division: the genus *Haplochromis*, comprising several hundreds of species.*

However, the challenge for taxonomists is, on the basis of solid findings, to arrive at a more natural classification than the one most recently published. The rejection of a revision must be argued in the form of a new revision, in the hope that, over time, the taxonomic classification will represent an increasingly accurate reflection of evolutionary history or phylogeny. This was what we were attempting to do. One by one we examined the right to existence of each of the genera specified by Greenwood.[18,19] Not a single morphological criterion has yet been found to justify the dividing lines envisaged by Greenwood between the genera. Our job was hellish but not unusual. Most taxonomists spend the greater part of their lives chasing other people's balloons: they live for them, puncture them, and release new ones.

A Fanning Out of Forms

How, without using reproductions, could I ever make clear to a Sukuma who had never seen a still life by Morandi that each of this painter's canvases was worth experiencing? Could these still lifes be described without giving the impression that they were mere replicas of each other? Was it possible to convince someone that the jars and bottles in Morandi's paintings were imbued with life? Fortunately for the Sukuma, I never attempted to do this, but right now I am faced with a similar problem. How can I introduce to the reader more than three hundred species of closely related furu that at first glance appear to resemble each other but that consistently reveal differences in shape and color?

Anyone who works with fish more intensively than with people over a period of years starts to see different personalities in the various species. The algae-eating philosopher, the snail-swallowing pimp, the larvae-sifting

*In addition to the genus *Haplochromis,* four so-called *monotypic* genera were distinguished, each of which includes only one species.

housewife ... it was with them that I worked for days on end. I projected human qualities onto the fish—an activity considered anathema by every right-minded scientist—and, in turn, saw fish in humans. There was no longer any distinction between the two. While shopping in the market in Mwanza I continually spotted traces of fish personalities in the faces of Africans, Indians, and Europeans. In passing, I whispered their Latin names under my breath, occasionally coming across someone without a double, one of those rare faces without an underwater counterpart. Wide mouths, protruding lower jaws, or, by contrast, small, short, highly underdeveloped ones; shrunken mouths or parrotlike beaks; long protruding teeth or a painfully misformed set of bicuspids; lips of all sizes, from tight, thin ones to thick, fleshy ones; round, straight and receding foreheads; a glazed stare fixed in large, bulging eyes; heavy jaw muscles embedded in a corpulent face. I spent most of my time studying the characteristic shapes of the fish so that eventually, however deadening this can be for visual alertness, I couldn't stop looking comparatively. Discovering similarities between the different species made me feel good. Identifying differences did the same. All that feeling good became almost too much for me.

An organism must assert itself actively in its environment and it is to this basic fact that researchers cling in their determination to classify. For this reason cichlids are classified according to the food they eat or the technique they use to digest their food.[20] We, too, used this relatively arbitrary system.

What we collected from the water:

Mud-biters These feed off organic waste, or detritus, on the lake bottom. Everywhere in the lake, at depths varying from two to more than thirty meters, masses of detritus-eaters are found just above the lake floor. There are at least thirteen different species. In calm weather they swim through a slowly falling shower of diatoms and blue-green algae, which ensures variety in their feeding habits.

Algae-scrapers At least three species are specialized in scraping algae from the rocks. They are always found along rocky stretches of shoreline or near the countless rocky islands in the lake. Looking into the mouths of these algae-scrapers, one sees a file in both the upper and lower jaws, each of which consists of several rows of small equal-sized teeth. Other species scrape filamentous algae from rocks with fewer, but sharper and

Figure 2.1
Algae-scraper.

Figure 2.2
Snail-sheller.

longer, teeth. The rock-frequenting fish of Lake Victoria are less differentiated than those of the older communities found in Lakes Tanganyika and Malawi. These lakes are also inhabited by species that rasp algae—finely or roughly—or bite them off. Moreover, there are species with long teeth that comb or quickly and roughly brush filamentous algae. Because of the differences in the anatomical grazing equipment and the feeding techniques, each species lives off a different diet.

Several species of algae-eaters are always found near vegetation. These species feed off algae that grow on the stems or leaves of aquatic plants.

Figure 2.3
Zooplankton-eater.

They are seldom found among the rock-grazers, just as the grazers are seldom seen near vegetation. These are not the only relatively immobile species. Most species are very much restricted to their traditional feeding grounds.

Leaf-choppers The foraging behavior of these species has never been observed, but the contents of their stomachs and intestines suggest that they chop or tear off pieces of macrophytes.

Snail-crushers Nine of these species are known to be specialized in breaking open snail shells. An extra set of "jaws" in their throats is heavily reinforced, being filled with teeth like millstones and powered by strong muscles. This pharyngeal jaw apparatus functions as a kind of nutcracker. In addition to the genus *Haplochromis*, this group also includes the genus *Astatoreochromis,* a snail-crusher prevalent throughout East Africa.

Astatoreochromis had been introduced in several places in Africa—including Cameroon and Zanzibar—in irrigation canals and rice paddies, in the hope it would decimate the snails responsible for transmitting bilharzia. Moreover, it was hoped that the people would be able to eat the fish. But this biological method of combating bilharzia proved ineffective as these fish eat snails only when nothing else is available. As long as soft, energy-rich prey such as midge larvae are present in sufficient numbers—and to our knowledge, this is always the case—these easily digestible prey are preferred.

Snail-shellers Cracking snails with reinforced pharyngeal jaws is not the only way to process this food source. At least twelve species approach and deal with snails in a different manner. The pharyngeal jaw apparatus of these species is not built to produce the crushing power needed to

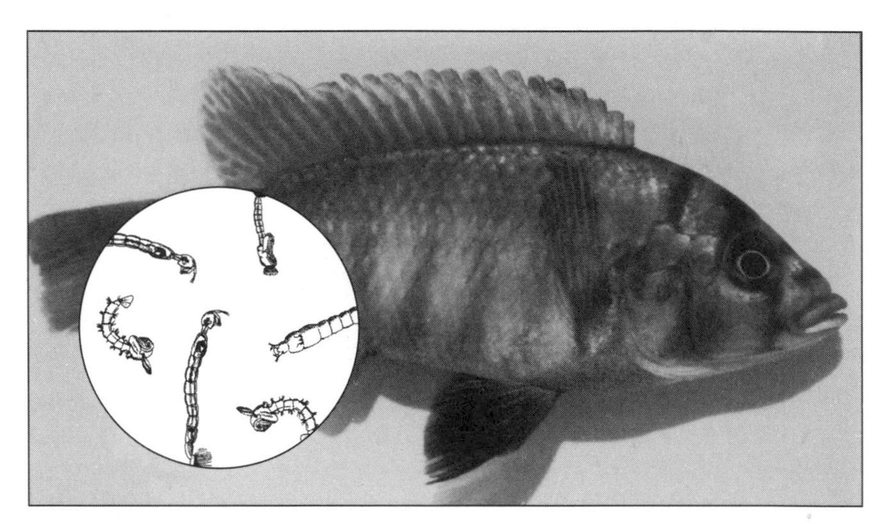

Figure 2.4
Insect-eater.

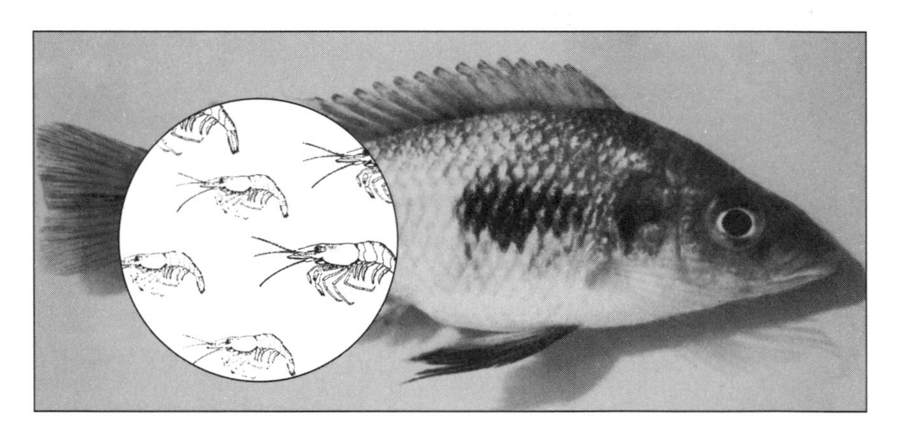

Figure 2.5
Prawn-eater.

crack thick shells. Instead, these species descend quickly on the slow-crawling snail and attempt, with their long, curved teeth, to seize its soft parts before it retreats into its shell. They shake the snail back and forth vigorously and push it against the lake bottom before prying the meat out of the shell. It is striking that this group of snail-shellers includes four genera: the genus *Haplochromis*, and three other genera that occur only in Lake Victoria and comprise only one species each.

Figure 2.6
Fish-eater.

Zooplankton-eaters Twenty-one small, slender species of furu live off zooplankton. Zooplankton-eaters are found in all the lake's habitats, from shallow to deep. The buccal cavity of these species has protrusible upper jaws and tends to resemble an elongated cylinder. By rapidly protruding their mouths, these fish are able greatly to increase the size of their buccal cavity, whereby crustaceans can be sucked inward. There are masses of these sucking fish in the open waters of Lake Victoria. The bottom-bound snail-crushers and algae-scrapers would never be able to suck as efficiently as these zooplankton-eaters. In fish that rely on their biting capacity for survival, the feature of protrusible upper jaws has been sacrificed in favor of structures that enhance biting power (powerful biting muscles and wide, short jaws).

Insect-eaters The list of insect-eaters is long, comprising more than 29 species, but little can be said about the appearance of most of them. They are relatively large (7.5—13 cm), but rather nondescript in many of their anatomical features. They take a mouthful of mud and sift out the insect larvae. The mud is later expelled through the gill covers. An unusual type of insect-eater is *H. chilotes,* which frequents rocky habitats. It is immediately noticeable because of its exceptionally fleshy lips. *H. chilotes* rhythmically presses its cushioned lips against the substrate and sucks in insects. The thick lips therefore appear to function as a kind of suction pump—the precursor of the kitchen plunger.

Prawn-eaters This group of thirteen species feeds exclusively on prawns. The fish are relatively slender and have large eyes set so close together in the narrow head that they almost touch each other. Their main prey, the small prawn *Caridina nilotica,* is found just above the muddy bottom, from inshore areas to deeper water. Prawn-eaters inhabit primarily deeper waters.

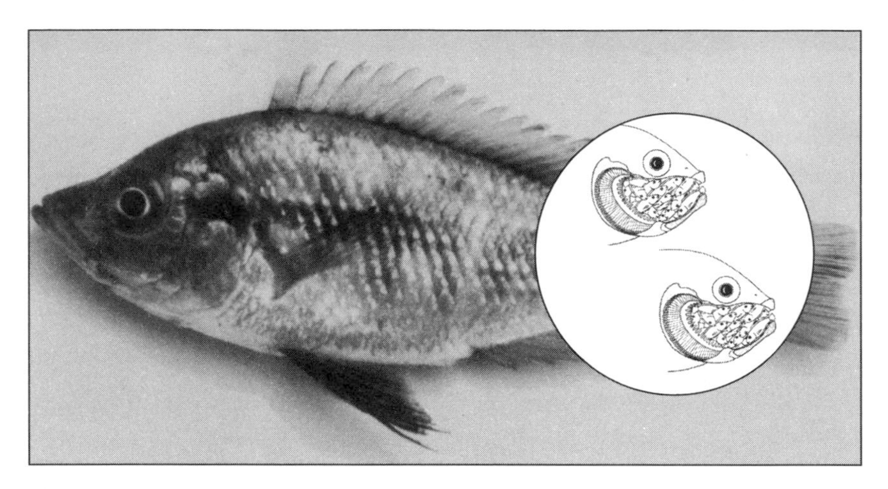

Figure 2.7
Pedophage.

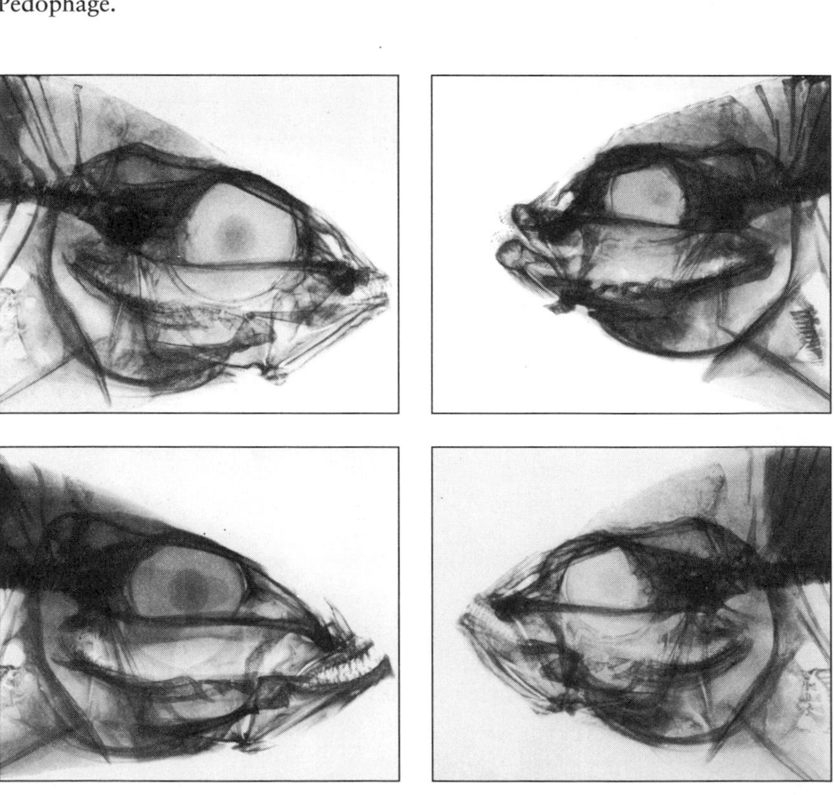

Figure 2.8a–d
X-rays of the heads of various species of cichlids taken from different East African lakes. It is the heads that have changed the most during the course of evolution.

Figure 2.9
Scale-scraper.

Fish-eaters The number of fish-eaters is exceptionally large: there are more than 130 species. These include species that feed on parts of other fish or their embryos. In the whole of Europe, there are only 129 different species of freshwater fish, belonging to 26 different families. This clearly demonstrates the uniqueness of the cichlid radiation in Lake Victoria. The number of fish-eaters alone approaches the total of all the freshwater fish in Europe. The fish-eaters that eat other fish in their entirety are roughly divided into two groups: those that lie in ambush or play dead and shoot out as soon as their prey approaches, and a second group of quicker, often streamlined fish that chase their prey and are known as pursuers. Strictly speaking, these fish-eaters have long, sharp teeth. The small teeth in the pharyngeal jaws are also sharp, so that the prey becomes filleted on its way to the esophagus.

Pedophages The fish-eaters also include 24 species of pedophages, or child-eaters, which feed on embryos or newly hatched fry. All the furu in Lake Victoria are mouthbrooders. The female swims around for weeks with her mouth full of eggs. These eggs later develop into fry. There is disagreement as to how the pedophages catch the embryos and the fry. At least four species of pedophages belonging to the genus *Cyrtocara* are found in Lake Malawi. The three species whose habits are known are all "rammers," but they each ram in a different way. One species approaches the female from below or behind and, when about two fish-lengths away from her, shoots toward her at a forty-five-degree angle, ramming her throat near the hyoid bone. Another species swims one-half to two meters beneath her, shoots upward, and then rams the underside of her mouth at

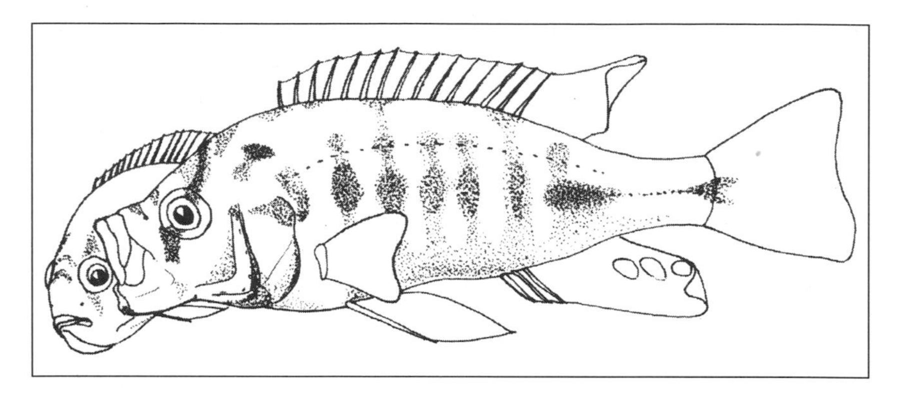

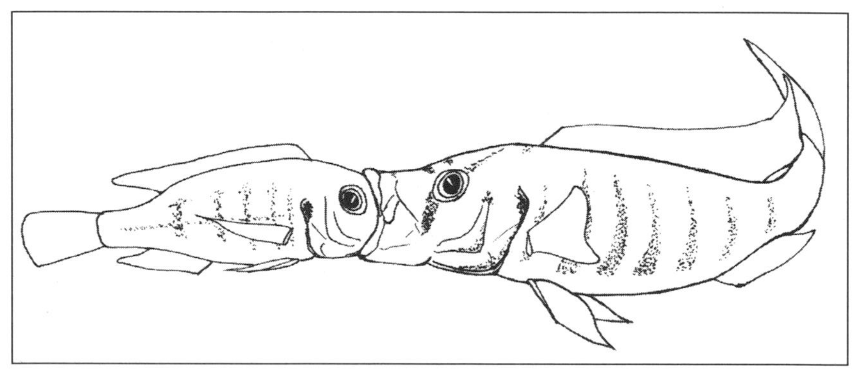

Figure 2.10
Snout-engulfing pedophage sucks empty the mouth of the brooding female.

a seventy- to ninety-degree angle. The third species approaches the female from above and crashes down, like a kamikaze pilot, onto the nose of its victim.[21]

Greenwood described a number of pedophages from Lake Victoria and noted the reduction in the number of teeth as well as their unusual implantation. A glance into the mouths of these fish suggests they would actually prefer not having teeth. Several rows of teeth have disappeared, although the front of the jaw still leaves room for some. They are strangely positioned, often curving outward instead of inward as one would expect given the function of teeth. Embedded in a layer of mucus, they bend forward so as to cause as little inconvenience as possible. In another of the pedophagous species, *H. maxillaris,* the remaining teeth have become even less innocuous by being permanently covered by inwardly curled lips.

Why is it significant that these fish don't bite? Greenwood believed that these species caught the brood by engulfing the snout of the female and subsequently sucking as hard as they could until it was empty. Sharp, inwardly protruding teeth would be fatal to fish that acquire their food in this bizarre way. Snout-engulfers with sharp teeth would run the risk of getting stuck on their victim or of being unable to disengage themselves quickly enough or even at all. "Unlikely" was Fryer's response to Greenwood's hypothesis: *H. barbarae* have also been found with embryos in their stomachs, but this species has normal, sharp teeth and therefore lacks the anatomical deformations found in other pedophages.[8]

The zoologist Wilhelm observed pedophages in aquaria and saw how one pedophage, *H. "rostrodon,"** engulfed the snout of a brooding female and sucked out the contents.[22] So Greenwood might have been right. Snout-engulfing does occur, at least in artificial environments.

What remained to be solved was how *H. barbarae* ended up with embryos in their stomachs—until it was observed that they lie patiently in wait during the courtship of a pair of cichlids from another species. Cichlids do not lay their eggs all at once but in small portions. The moment a batch of eggs are laid, *H. barbarae* shoots toward them like a bolt of lightning, managing to snatch up some of them before the female has the chance to take them into her mouth. In this case, a prediction about feeding behavior based on anatomical research—that is, that this species could not be a snout-engulfer—proved true.[23]

Scale-scrapers Another species found in Mwanza Gulf has the appearance of a fish-eater but the teeth of an algae-scraper. It has a long rasp, comprising eleven rows of teeth lined up for battle and loosely implanted, in both the upper and lower jaws. This species, *H. welcommei,* launches attacks on other fish and rasps scales off its victims. The scale-scraper is believed to approach its victim open-mouthed and to head right for the tail area, which is covered with smallish scales. The scales are not ripped out of the body one by one but scraped off roughly.

Scales are rich in protein: about as much energy is generated by a single scale as by one planktonic crustacean. Fortunately for both parties, scales regenerate. The scales are found tightly packed like rolls of coins in the stomachs and guts of scale-eaters. It is assumed that these scale-eaters descended from the algae-scrapers of the rocky habitats, following an adaptation of their eating habits. Overpopulation of the rocky areas is thought to have induced them to seek new food sources. Such changes in behavior often lead to the development of anatomical adaptations. In this case, the fish developed from reasonably plump, rock-frequenting specimens into more streamlined, agile hunters.

*We assigned a nickname to the species that were not already scientifically documented. The nickname is indicated here in quotation marks.

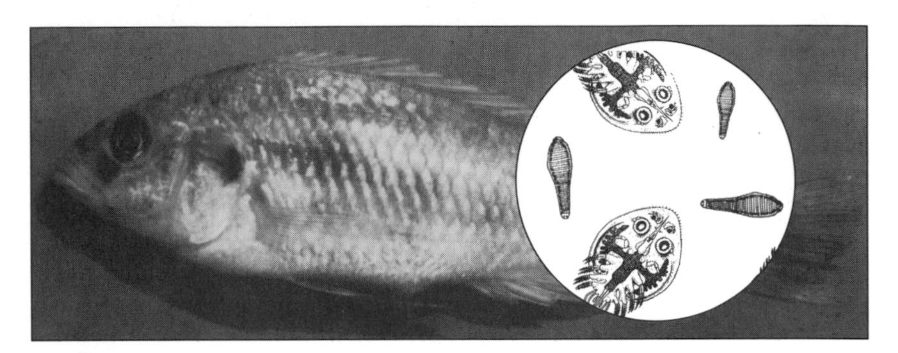

Figure 2.11
Cleaner.

Cleaners Two species in Mwanza Gulf have remarkable feeding habits. Their stomachs and intestines are full of pleated, transparent raincoats. Two small wheels are mounted side by side on each raincoat. The raincoats are the carapaces of parasitic crustaceans (argulids), and the mysterious wheels are the suction cups used by these ectoparasites to attach themselves to their host.[24] The analogy with biotic communities of coral reefs becomes increasingly evident. Coral reefs, too, house widely differentiated fish communities, although their inhabitants belong to distinct families. Cleaners—fish that remove parasites from other fish—have been sighted there as well.

Once, while assigning names to the different species of a catch, I found myself looking into the eye socket of a fish, a gaping hole: the eyeball had disappeared. Was this the work of the legendary eye-biter? Fishermen from Lake Malawi told us stories about a fish, *H. compressiceps,* that was supposed to be specialized in plucking out eyes, though it also consumes other food. Why, in a lake already populated with snout-engulfers, scale-scrapers, cleaners, snail-shellers, and leaf-choppers, could there not also be room for eye-biters? The story about the eye-biter is probably a myth. No cichlids have ever been found in any of the East African lakes with a significant number of eyes in their stomachs.

But all the other food specialists were found in Lake Victoria. In terms of complexity, the system was in no way inferior to the spectacular terrestrial African ecosystems of the Serengeti plain or the Ngorongoro crater. It would be all the more fascinating if these hundreds of food specialists all

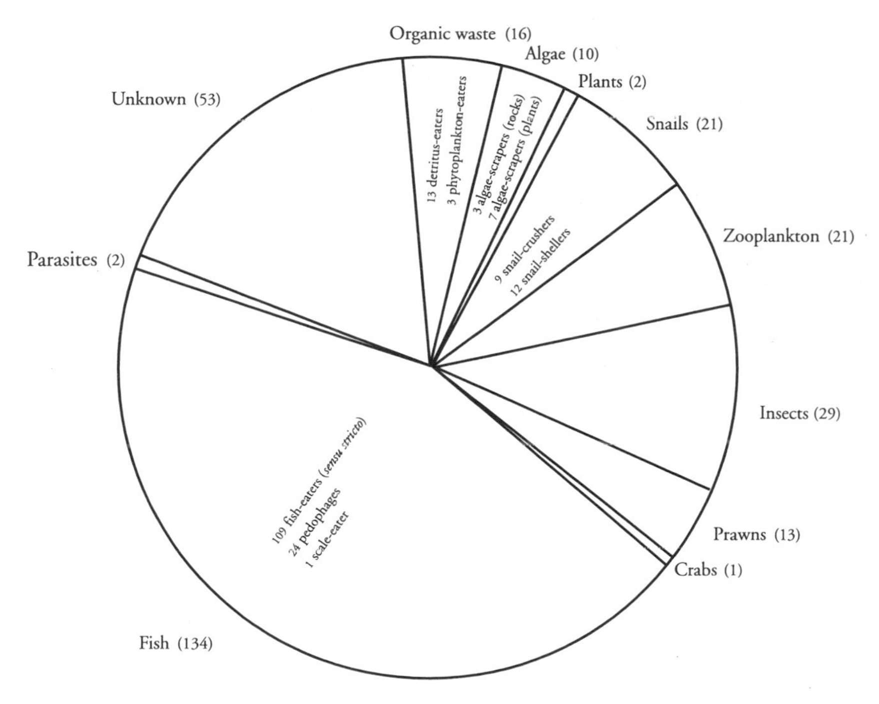

Figure 2.12
Fish-eaters, snail-eaters, insect-eaters, algae-eaters, zooplankton-eaters, mud-biters, and other trophic groups belonging to the species flock of Lake Victoria.

stemmed from a single riverine ancestor that had eventually found its way into this lake. Only if that were the case could the furu satisfy the definition of a species flock. But could this be determined? (It would be, at a later date.) By comparing the DNA of the different furu species of Lake Victoria with each other and with the DNA taken from cichlids from other lakes, the extent of the similarities could be established. Were we dealing with a species flock? Only if that were certain would it make sense to draw up a phylogenetic tree, the basic structure for describing the evolutionary history of the flock. How this fanning out of forms evolved, and on the basis of which evolutionary principles, these were matters for later.

3

Maji moto: Furu Forebears

The sun was already high overhead when I arrived at the airport and parked the car. People were standing waiting at the door of the arrivals hall. I asked a porter the time. The clock in the hall had stopped and I hadn't owned a watch for months. At Elimo's request, I had traded it with him for a rooster and a *sonzo*, a traditional Sukuma basket. These watertight baskets were woven by women and used at celebrations as vessels for consuming vast quantities of *pombe*, a millet beer that tasted like bitter vomit. Blades of grass dyed with manganese were woven into the baskets in geometric patterns with a symbolic significance. It wasn't always possible to find out what the patterns meant because the arrival of the *mazabethi*—the aluminum dishes named after Queen Elizabeth that had been introduced on a large scale under British rule—had signified the end of the *masonzo* culture. I spoke to an old woman in a little village who, after more than thirty years, was still incensed about the mazabethi. She told me that the symbol—composed of triangles—on the basket I had traded with Elimo meant *kitenge,* or wraparound.

"My husband dresses his first wife well, but I, one of his younger wives, have to wear a threadbare wraparound."

Sukuma women were not allowed to air their complaints verbally, but they could vent their frustrations in this symbolic form.

"The arrival of the mazabethi spoiled it for us," the old woman continued. "*Sisi wanawake*, we women, we used to weave baskets while sitting around and chatting with each other. I don't see anything wrong with that. Each woman did her best to make the most beautiful basket possible. The mazabethi put an end to all that."

I was forever on the lookout for baskets bearing symbols that might relate to fish, but I hadn't come across any yet. In the more than one hundred baskets I had seen up till then, there were very few decorated with symbols in any way related to the lake. One symbol stood for "the corner of the lake," another for "floating papyrus island." But whoever became acquainted with the Sukuma solely through the symbols on their baskets would easily get the impression that their world ended at the lakeshore. The lake, the edge of the world, a watering place for cattle. A nation of farmers and cattle-herders, not of fishermen.

Rumor had it in Nyegezi that a plane would land that day. In itself such a rumor had little significance, but for several days I had been expecting the molecular biologist S., who was coming from America specifically to collect DNA from the furu, and I was doing my best to meet him. He was going to try to establish whether the many hundreds of species of furu in the Victoria basin really descended from a single ancestral species, a species that had at one time—albeit less than a million years ago—found its way into the virgin lake. Was it a species flock that had emerged over a short period of time as a result of prolific speciation, a monophyletic group? Or did the Adams and Eves of the furu from this lake belong to different species, a polyphyletic group, like the cichlids of Lake Tanganyika?[25]

Perhaps molecular biologists would eventually succeed in establishing a phylogenetic tree for the furu based on molecular features. If that coincided with a phylogenetic tree based on morphological features, then faith in its validity would increase considerably. A verified phylogenetic tree. Unverified phylogenetic trees sometimes say more about the features their makers consider important than about the evolution of the group of organisms to which they refer.

S.'s letters had given me the impression that he was irresistibly attracted to the furu's DNA. This might prove inspiring. We both felt that Greenwood was right when he wrote: "The cichlid species flocks are microcosms, repeating on a small and appreciable scale the patterns and mechanisms of vertebrate evolution."[13] This created a bond. But now I'd had enough of waiting. Hurry up and land. I flopped down in a chair in the arrivals hall opposite three enormous Arab women squeezed closely together on a little bench. There wasn't a European in sight, only Tanzanians, Arabs, and Indians.

An athletically built girl in plimsolls announced that it might be several hours before the plane arrived. A large fan whirred furiously next to her.

Her hands flew in all directions in an effort to hold down her flimsy skirt—like Marilyn Monroe standing above the subway vent. She announced that the plane had been rerouted at the last moment. How could it be that all the wanderers knew this except me? Did they inquire by telephone about arrival times? I didn't have a telephone.

I bought a can of mango juice in the arrivals hall, sat down on a wooden bench, and began preparing myself for the arrival of our guest. I pulled out an article by Melanie Stiassny, curator of the American Museum of Natural History in New York. The last time I had seen her she had been repairing the damaged shoulder of a herring that had found its ultimate destination in one of Joseph Beuys's assemblages. "Mmmm, herring," I mumbled banally to the three Arab women. They turned their heads away from me in one fluid movement, like a three-headed monster.

Let me concentrate, though, on Stiassny's article on the phylogeny of the family Cichlidae. She reported the following: the family has a wide distribution. Cichlids occur in the rivers and lakes of South and Central America, Africa, and Madagascar. As for their distribution, they are limited mainly to fresh water, but the odd genus is found in the brackish coastal waters of southern India and Sri Lanka.[26] On the basis of comparative anatomical research, Stiassny came to the conclusion that the family Cichlidae is a natural group, in other words, has descended from a common ancestor.[27]

Stiassny, too, was interested in drawing up phylogenetic trees based on anatomical features. The aim was to create a natural tree. Such a tree provides information about the chronology of speciation events within a particular group of organisms, about the background of the radiation, and about the differentiation into different ecological niches. An important point was whether, within certain habitats in the same lake, for example in a habitat with a sandy bottom, mini-radiations of different food specialists could develop. Did trophic types, such as the phytoplankton-eaters, zooplankton-eaters, snail-crushers, and fish-eaters, evolve over and over again in different habitats? Or were the closest relatives of species of one trophic type always different species belonging to the same trophic type?

Without precise knowledge of the evolutionary history of all the species of a particular lake, it is impossible to determine which similarities between the fish should be attributed to a common origin and which are the result of convergent evolution caused by exposure to identical environmental factors. A classic example of convergent evolution is whales and fishes, which, despite the extensive similarity of their external features, are in no way related. But convergent evolution might also be important in the case of the furu in Lake Victoria. It has been established that the cichlids from the different East African lakes evolved convergently: one group of algae-grazing species occurs in each of the three Great Lakes—Tanganyika, Malawi, and Victoria. The species resemble each other closely in appearance. They are deep-bodied and have wide heads with a highly curved profile and a scraping apparatus similar to a rasp. Geneticists have compared proteins isolated from the tissue of these cichlids by means of gel electrophoresis, a technique for separating proteins according to molecular size.[28] Such comparisons of species from Lake Malawi revealed minute genetic differences. The same applied to the furu of Lake Victoria. Species belonging to the same flock displayed barely diagnostic genes. They could only be distinguished from each other on the basis of frequencies of certain genes, a powerful indication of the recent origin of these flocks. But there were considerable genetic differences between the algae-scraping species from Lakes Malawi and Tanganyika. In both these lakes, the specialized morphological structure and the shape of the skull and teeth of the algae-scraping species originated independently. These species can therefore be seen as an example of convergent evolution. Later research confirmed that the species flocks of the three African Great Lakes originated independently.[29] But the completion of natural phylogenetic trees is still a long way off.

Unfortunately, it was much too hot a day to read Stiassny's sound but extremely detailed studies. The article slid further and further down my knee as I dozed to the soporific whirring of the fan.

The athletic girl made another announcement: there was a technical problem with the aircraft, which was still grounded in Dar-es-Salaam. I stood up, swore under my breath, nodded in the direction of the three Arab women, and left the hall. A group of Africans was standing next to the Land Rover in the hope they could get a lift back to town. There was no public transport running at the time. I drove back to Mwanza with my Land Rover full of hitchhikers.

The town was dozing. It was the midday break. Shops were closed and the windows covered with wooden shutters. The doors and shutters in turn were protected by gratings secured with heavy padlocks. Iron on top of wood on top of glass. The shopkeepers cherished their wares like treasures.

I drove slowly down the main street, trying to decide how to make the most of my day. We needed floats for the gill nets, but I couldn't buy them at that hour. Only when the ninth hour of the day had elapsed would the shops reopen. The Tanzanian day begins at six o'clock in the morning. That is the first hour of the day. It ends twelve hours later, when evening sets in. A by-product of this manner of keeping time is that Tanzanians and Western tourists forever miss each other—with mathematical precision. Tourists systematically show up for appointments six hours too early. After several weeks of being out of phase with the inhabitants of this country, they return home disappointed, believing it is useless to try to make an appointment with a Tanzanian.

Incidentally, it is true that our predilection for punctuality strikes the majority of Tanzanians as overdone. Most of them prefer not to make an appointment but show up when it suits them. Meetings take place when the time is right. They are even inevitable. It is better not to try to arrange them. I was once again reminded of the extent to which our conventions of punctuality strike Tanzanians as strange when I received a letter from a Tanzanian friend. The letter began as follows: "This morning at exactly ten past four I found your letter in our mailbox ..."

The Tanzanian concept of time not only reduces the possibility of encounters between Tanzanians and tourists, it also creates confusion among the Westerners themselves. Just before the agreed-upon time, doubts begin to arise. Might the other person already have liberated himself from the straightjacket of the Western clock? Does he think like a white or black person? Was the appointment made in Swahili or a Western language? I contemplated whether this might not be why the American was nowhere to be found.

During the past few days, I had driven back and forth four times from the institute at Nyegezi to the airport via the city—to no avail, and using valuable fuel destined for the research boat. It seemed pointless to consume several liters of gasoline in returning to the institute only to have to come back to the airport for a fifth time in the late afternoon. I recalled the many hours I had spent in the waiting room of the government office that issued gasoline rations; how I had had to explain to a friendly civil servant the intended destination of the gasoline and the amount of gasoline I would be using and for what purpose; and the relief I had felt when he had deemed my intentions "gasoline-worthy" and written out a *kibali,* or receipt, good for several dozens of liters of fuel.

Receipt in hand I had skipped down the steps of the government office and driven straight to the designated gas station—only to have to wait there in an interminably long line. The gasoline, a particle-filled sediment that had to be sucked from the bottom of the tank beneath the station, sputtered its way to the surface. This took time. Those who were waiting chatted with each other or dozed behind their steering wheels, waking with a jolt every time the line advanced a few meters. One man with a jerry can tried surreptitiously to cut into the line. When he was within thirty meters of the pump, he suddenly hastened his pace, ran up to the attendant, and fell onto his knees, begging for gasoline. The attendant, a slender Nilote, pulled the nozzle out of the tank that he was filling at the moment and moved it slowly toward the jerry can. People in the line who saw this happening stalked forward indignantly, grabbed hold of the man with the jerry can, told him off, and pushed him away from the pump. The pump attendant stood there looking fearful and distraught.

"This man does not know how to handle the case," said the Tanzanian who was standing in front of me in the line, adding, in an almost apologetic tone: "*We* are accustomed to shortages, but it must be difficult for you expatriates."

"But that's the charm of this country," I had replied. "Besides, we can always leave when we've had enough."

While watching the incident at the pump, I had muttered incantations to the effect that the gasoline wouldn't run out before it was my turn, that there wouldn't be a power failure, that the pump wouldn't close before I got my ration, and that I wouldn't become one of the countless moaning wanderers of Mwanza.

We Need Business, Brother

I waited in town until the next plane arrived, parking the Land Rover in front of the New Mwanza Hotel. I removed a notebook from my bag and began writing: "S. didn't arrive again. Not getting anything done. Why and for whom am I actually here? For the Tanzanians, for the biologists in Leiden, or so I can receive guests? Have I come to live on the Equator so I can line up for gasoline two days a week? So I can spend one day a week looking for food in order to have enough energy to line up for gasoline? Everyone is desperate here. Even the moles are suffering, under the weight of the tarmac runway, which at several places looks as if one of them is trying in vain to stick its head up through it. I've no doubt that many exciting things are happening underwater, but when will I ever get there? For God's sake, as long as there's no gasoline, write something about Mwanza. Stop what you're doing, immediately. Two officers are sitting on the terrace of the New Mwanza Hotel. They're watching me." I put my notebook away again and peered at the terrace where the civil servants were filling their bodily tanks: two rotund men with surly, snail-crusher heads. They were accompanied by two budding, attractive young women, already rather plump, poised to expand during the next few years. Their faces gave the impression of being heavily made up, but this might not have been the case as I hadn't seen any make-up for so long. Women seldom wore make-up here, so a touch of it was enough to stimulate the senses, at least mine. Both women were clad in worldly dresses, not the printed wraparounds normally worn by peasant women. They searched nervously in their handbags, the permanent accessory of the local whore. Come now, you don't have the courage, I muttered to myself. Oh yes, you bought a book at the evangelical bookshop, *The Prostitute in African Literature*.

A tawny-skinned waiter in black trousers and a white shirt was standing near the terrace entrance. Looking bored, he gazed steadily past the guests until one of the men stood up and passed him as he left the terrace. The waiter took a step back with his left leg as elegantly as a toreador who dodges an approaching bull at the last moment. Then he quickly pulled back his right leg while swishing a dangling tray upward past his chest. The man passed by him, taking no notice of him as he did so.

I climbed out of the Land Rover. As I locked the door, the corpulent giant left the terrace. He was dressed in a khaki-colored suit—the Tanzanian equivalent of the Mao uniform—through which his plump light-brown flesh was clearly visible. He wore a leather string around his neck from which hung a disc-shaped white amulet. A gold watch glittered on his left wrist. As he stepped off the curb he almost lost his balance. I hastened toward him, but he dismissed me with a wave of the hand, adding: "I am on my own." Weaving slightly, he moved toward the Land Rover, pointing to the words written on it. When he was close enough, he made several attempts to pronounce the words, but the "ch" sound of the first word—"*Haplochromis*"—proved particularly insurmountable. When he was within arm's length of me, a strong familiar odor descended upon me. He emanated a heavy, rancid odor that, every time he spoke, was accompanied by the sweet smell of *conyagi*, a powerful distillate. Fanning his face with a hotel napkin, he inquired gruffly: "Brother, what is this Haplo...?" He continued before I could reply: "We need trawlers, not this Haplo... We want commercial fisheries on Lake Victoria. Remember, you are in the Third World now. We need business, brother. I want to see you in my office." He turned away abruptly, muttering: "But not today. I need another hot tea ... Thank you so much for this conversation," then waddled back to his companions, whose gaze had been fixed upon us steadily all this time.

What possible significance could my presence here have for the Tanzanians? I used the last of their gasoline in pursuit of a guest. But let me address myself on a more encouraging note: as long as a can of Portuguese sardines was better protected here than the rarest fish in the lake, perhaps there was no harm in my being here. Maybe I was even doing something to protect organisms in which no one else was interested. Besides, it was very likely that the project would be expanded to include a practical component involving research that would be of immediate benefit to the Tanzanian people. Indeed, someone had already flown over from the land of Lorentz and Leeghwater: practically oriented Melle, who was never at a loss and who was capable of doing anything as long as it served a purpose. He might even gain access to a seaworthy ship that was being built in the Netherlands.

But why didn't we go home if the Tanzanians who were in charge saw no point in our work? Sometimes I had the impression that they thought: Oh no, there's another group of Dutchmen. If you don't keep your eye on them, they'll start "do-gooding" and, before you know it, they'll want a declaration in sextuplicate saying that that's what we've been waiting for all these years.

As I walked along, I glanced into a hairdresser's shop that was about to close. A woman with braids shooting out of her head in all directions took one last look in the mirror before leaving the shop. After she had stepped out on the street, carefully feeling where her hairdo ended and the outside world began, the owner closed the shop and a series of painted heads appeared on the light blue shutters—the range of hairdos from which you could choose: "James Brown," "Disco," "Nairobi." All those laughing heads with their baroque hairdos, floating in a blue sky, seemed to be speaking to me encouragingly: "*Hamna wasi wasi*, don't worry, wanderer. Go ahead and catch all those fish. After a few years you'll leave for home and we'll inherit a boat that we'll be able to use for years."

If only S. were here. What would he look like? Molecular biologists were usually easily recognizable: they were well fed and always carried a little plastic drum filled with liquid nitrogen. They nurtured the drum like a bambino. Strangers were not allowed to approach it. Whoever ignored the warnings and touched its contents when no one was looking burned his fingers severely. Organic tissue destined for DNA research could be preserved for months on end in liquid nitrogen. The molecular biologists asked us which species they should use for establishing a phylogenetic tree, because they had no idea what the organisms they were studying looked like. I always found this strange. They, in turn, found it strange that I had absolutely no idea what genes looked like. Surely it was the content that mattered, not the form. After consultations with us, they placed their order: eight purple spots, seven parrot beaks, ten tridents, nine flaming heads ... As soon as their drum was full, they would depart again. Having returned to their modern laboratories in the West, they would try to establish the phylogenetic relationships between the species whose DNA they had collected.

I sat down on a rock next to a group of boys. They were making little oil lamps from old cans, to the sounds of hammering, tapping, and radio music. In order to get more out of S.'s visit, I reviewed the three-dimensional structure of DNA, detailed in Watson and Crick's rope-ladder model:

Every cell in the body of multicellular organisms has a nucleus. In the nucleus is a double set of chromosomes. The chromosomes consist of DNA, in which genetic information is stored. The three-dimensional structure of DNA can be visualized as a rope ladder consisting of two pieces of thick rope of equal length adjoined at regular intervals by rungs. Turn the rope ladder around on an imaginary axis. Do this several times, not too tightly, until it forms a double helix. This model, developed by Watson and Crick, gives a clear idea of the three-dimensional configuration of DNA.

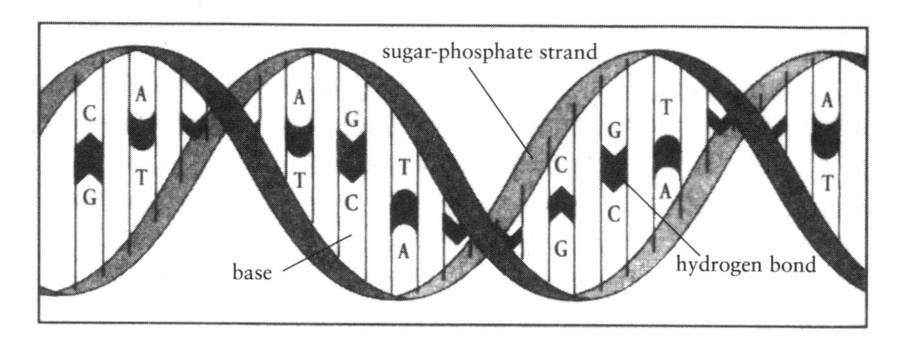

Figure 3.1
Graphic representation of the structure of DNA. A = adenine, T = thymine, C = cytosine, G = guanine.

Each of the two spiraling DNA strands is composed of rows of bases, the nucleotides, lined up in sequence. There are four bases in nuclear DNA: adenine (A), cytosine (C), guanine, (G), and thymine (T). Adenine in one strand is always aligned with thymine in the other, in the same way as cytosine is always matched with guanine. If the base adenine is known to be located at a certain position on one strand, the base thymine will always be located opposite it in the complementary strand. In the rope ladder model, one rung of the ladder symbolizes two complementary bases joined together by a hydrogen bond, while one rope stands for a single strand, comprising a series of phosphate groups and sugar molecules.

When all cell divisions take place without error, the genetic information stored in the nucleus is identical in each body cell. When a cell is about to divide, the two complementary strands of DNA separate. Before cell division has been completed, the two relatively unstable hydrogen bonds between the complementary bases are broken down and the double helix separates into two single strands. Each of these strands forms a duplex with a complementary strand, according to the principles of base pairing. Eventually, a newly formed double helix appears in each of the daughter cells, so that each contains DNA with exactly the same base sequence as the parent cell. The DNA codes for proteins, which comprise amino acids. Complex proteins acquire their characteristic structure and properties depending on the composition of the amino acids.

A sequence of three bases, a triplet, indicates which one of the total of twenty existing amino acids is to be used. Some triplets code for the same amino acid. This process, in which amino acids are joined together, would continue in perpetuity if the code did not indicate when a certain protein was completed. Three triplets do not code for the synthesis of an amino acid but indicate when a protein is completed.

The formation of amino acids and their assembling to form a protein takes place not in the nucleus but in the surrounding cytoplasm. The nucleus contains the scenario for protein synthesis but the actual synthesis takes place in the cytoplasm. The information, encoded in the nucleotide sequence of the nuclear DNA, gets translated into messenger RNA according to the principles of complementary base pairing and is then transported to the cytoplasm. The cytoplasm contains ribosomes, small organelles comprising RNA and proteins. These play a key role in protein synthesis. As soon as the ribosomes start to affect the messenger RNA, they begin to assemble amino acids, according to the specified code—into proteins. A special form of RNA, transport RNA, is responsible for bringing in the amino acids indicated by the code as necessary for the protein synthesis. Some proteins are absorbed into cell structures. Others, the enzymes, are responsible for catalyzing a large number of metabolic processes.

The genetic code is virtually identical in all living organisms. Regardless of the species, a specific sequence of three nucleotides will always be translated into the same amino acid.[30,*]

Well, let him come, this gene freak with his little drum. I continued walking. At the foot of a large tree, indestructible sandals were being carved out of old car tires. I removed my shoes, tried on some sandals, and purchased a pair. Several men looked at my white feet with amusement, as if they were exotic animals. Were these animals going to do something or just keep standing there? They continued walking. There was a limited selection of wares: bars of soap, aluminum dishes, and cups; papayas, mangos, and limes; bread, tea, *mandazi*, and little rice-meal pancakes. There was also a pile of the *Daily News*, the state-controlled newspaper of the only legal party, the *Chama cha Mapinduzi*, the Party of the Revolution. When paper was in short supply, the *Daily News* sometimes didn't appear for a week, but if at all possible, it was printed.

A traditional healer beckoned to me and pointed earnestly to a sign bearing the names of the illnesses he treated: headaches, measles, malaria, tuberculosis, and many others. I was unfamiliar with many of the Swahili names but I did recognize an impressive number of venereal diseases. It is strange to imagine that the reason sexual reproduction came into existence was to resist illness. This may in fact be the only justification for the existence of the male sex. Sex produces new genetic variations. The genetic material of the parents gets reshuffled. Among the resulting variations, individuals are found occasionally that are capable of resisting pathogenic organisms. Organisms that reproduce asexually produce far fewer new genetic variations and therefore have a much greater chance of being decimated. But when sex itself eventually came into being, it too became an inexhaustible source of disease.

*See *De DNA-makers. Architecten van het leven*. H. Schellekens (ed.). Natuur & Techniek, Maastricht, 1993. The factual information on DNA was taken mainly from this book.

I rummaged through the wares on the table. There was a great assortment: little pieces of wood, small bags with powder, jars with ointments, snail shells in a wooden dish, pieces of skin from civet cats, quills from porcupines, and copper bracelets thought to have a protective effect. There was also a *lupingu*, a triangular piece of shell carved from the shell of a sea snail. During the German occupation at the beginning of the century, a lupingu had been worth a great deal: four lupingu were equivalent to one goat, an aged Tanzanian had once told me. It had been difficult for the Sukuma, who lived thousands of kilometers inland, to get their hands on them. During celebrations, marriageable girls would walk around wearing as many lupingu as possible around their necks. This was to give potentially interested suitors an idea of the size of the dowry the parents expected if they were to consider giving away their daughter. Under English rule, when lupingu were imported in vast quantities, their value plummeted. "What a life, father," said the healer. I nodded approvingly and said farewell.

A heartrending shriek brought my reverie to an abrupt end. The sound emanated from a large loudspeaker in the pure white minaret of a mosque. A group of crows dozing on the mosque wall fled in panic. Two cattle egrets perched in a tree opposite readied themselves for flight—stretching their legs and flapping their wings—but then settled again. Meanwhile, the sound welled up as if the singer were being castrated, before becoming a penetrating wail that cut through you like a knife ... Has the genetic code been unraveled? Sincere congratulations for your newfound knowledge, you misguided nit-pickers. But don't for one moment think that the deciphering of the complex mechanisms of nature in any way diminishes the Great Mystery of Existence ... Did I understand it correctly or was I distilling the wrong message from this wailing lament? It became deathly still. A solitary beggar was the only one still standing at the entrance of the mosque. His swollen legs were covered with unsightly sores. He hummed softly, accompanying himself by hitting two coins rhythmically against each other. An emaciated cat crossed the white-tiled patio of the mosque, which was covered with shoes and sandals. It followed a zigzag course, moving from one scent to the next.

I walked back past the market toward the lake. Patel, Al Salim, Abdel Hussein. The names on the shopfronts sounded Indian or Arabic, but not

African. Should I go to the post office and see if there was a message from S.? If it wasn't really necessary, I'd prefer not to. The sight of an empty mailbox had become unbearable.

No love letter, no letter from friends, no news from the molecular biologist, only a brown envelope from Leiden University. A notice from the household unit of the biology department. The coffee break would henceforth begin twenty minutes later. Please follow this guideline, which is necessary because of the overloaded schedule of the canteen personnel. End of notice. I decided to buy some stamps after all.

"*Hamna*, they're all gone," said the girl behind the wicket.

"No stamps?"

"*Hamna*," she repeated, showing me the empty pages of her stamp book.

There were still stamps available in the lowest denomination, but I would need so many of them to send a letter to Europe that the girl's supply would be totally depleted. I thought about the letters I'd sent previously. First, the envelopes had become progressively filled with stamps, as the stamps of the higher denomination became scarcer. By the time there were only stamps of the lowest denomination, I had attached wings to my letters to create the necessary space. Short letters with a single wing and long ones with double wings. I had started enjoying it and wrote one letter after another just so I could dispatch another airborne double-decker.

"Can I give you my letters, pay the postage, and send them without stamps?" I asked the girl.

"That's impossible," she said, looking glumly past me. "A letter must have a stamp. That's all there is to it."

I nodded in agreement and was about to leave when she pushed her hand through the wicket, beckoned to me, and said: "Wanderer, wanderer, you can go to Shinyanga. My sister told me they still have stamps there, including those of the highest denomination."

"But that's a day's travel by bus," I said.

"So what?" replied the girl laughingly. I thanked her for her advice and left.

At the office of the national airline company I was told the plane's technical problem hadn't yet been solved. It was the only aircraft available for domestic flights. *Nimeshindwa*, I'm beaten, I mumbled to myself. I wanted a cigarette, to smoke, immediately.

A cigarette. I wasn't the only one who wanted one. They weren't easy to obtain. They'd been sold out in the shops for months already. But perhaps I could find some on the street. They were sold on the black market. Outside a cinema I came across some boys selling Sportsman cigarettes individually. I hesitated. Should I smoke a cigarette? Perhaps it would be better to get drunk. The boys thought I was worrying about the price. I didn't have to buy a whole cigarette if it was too expensive. I could also just take a puff. A shilling a puff, and they would count the puffs. I treated myself to a whole cigarette and savored it while sitting on the curb. I resolved to put S. out of my mind once and for all. He would never leave Dar-es-Salaam.

Then the boy with the package of cigarettes asked if I would buy a cigarette for him.

"But they're yours, aren't they? Or do you want to smoke one yourself?"

No, he wanted to sell them.

"So you're doubling the price of a cigarette?"

"Yes," he replied shyly, "then I earn more."

His reasoning was infallible. I bought the rest of the cigarettes in the package for double the price and divided them among the three boys. This caused great confusion. As they walked away, I heard them cursing and shouting, offering money for each other's cigarettes. Even small-scale help can disturb the peace and be detrimental to public health.

A week later I received a message from S. He apologized for having been unable to make it.

The Soldier and the Mattress

Again I was waiting at the airport for a biologist who was coming to collect DNA from the furu. While we had been lining up for gasoline, suffering from malaria, and fishing as if our lives depended on it, a breakthrough had been made in molecular biology in the West. A technique had been found whereby DNA could be traced and subsequently copied as often as necessary. The discovery had scarcely been made before molecular biologists started flying here again in droves. We still didn't know whether we were dealing with a genuine species flock. No good phylogenetic tree based on molecular features had yet been established.

Slowly changing nuclear DNA was not the most suitable substance for comparative research into young species such as the cichlids of Lake Victoria. The cytoplasm of the cells of vertebrates, however, contains small particles called mitochondria. These mitochondria contain DNA that *is* suitable for comparative research into young species. Mitochondrial DNA (mtDNA) is a circular, double-stranded DNA molecule. Each cell contains several hundred mitochondria.

Each mitochondrion contains five to ten copies of circular DNA. It is about sixteen to one hundred thousand base pairs long and contains genes that code for proteins—ribosomal RNAs—and genes that code for transport RNAs. It also contains parts that do not code for protein. These parts of the mtDNA are called the noncoding or control regions. In many organisms, the base sequence of all the mtDNA, which contains dozens of genes, has already been established.[29]

Mutations occur occasionally in DNA. In the simplest case, the mutation consists of the substitution of one base for another. Base pairs can be added or dropped or more radical reorganizations of larger segments of DNA can take place. Mutations occur unexpectedly, but it is generally accepted that there are certain limits to the frequency with which they occur in specific segments of DNA. When mutations in DNA occur with regular frequency, each mutation can be considered a tick of the "molecular clock." The extent to which the DNA of different species has diverged can be seen as an indication of the time period that has elapsed since barriers to reproduction between the species came into being.

If a comparative study were to be made of the base sequence of segments of nuclear DNA taken from such closely related species as the furu of Lake Victoria, it would reveal no differences. The base sequences would be identical. In such cases, mtDNA can be useful. It has been established in mammals that the mutation frequency of mtDNA is five to ten times higher than that of nuclear DNA. Moreover, within the circular mtDNA molecule, mutation frequency varies from region to region, reaching its highest level in the control region. Mutation frequency for segments of DNA from regions that do not code for protein is two to five times higher than for segments of DNA that do. This is why preference is given to using segments from the control region when doing comparative research on young species that differ little from each other.

Taxonomists attempt to establish phylogenetic relationships between species, thus slowly but surely expanding the phylogenetic tree based on molecular features. Genetic archives can be useful in this kind of work. Data concerning the base sequence of a segment of DNA can be sent to a genetic archive. There, the archive-documented organism with which the study object is most closely related can be determined. Several years ago, a research group that was attempting to establish the phylogenetic tree of a certain worm sent the relevant data to a genetic archive, requesting an indication of the most closely related organism. They were initially very surprised by the results. In genetic terms, the worms bore a striking resemblance to the domestic cow but were light years away from their fellow worms. After some thought, though, the researchers realized what had happened. The worms had been fed on beef's liver *ad libitum* and this was what their intestines had been filled with during their last meal, before they were slaughtered and subsequently minced for the sake of DNA research.

I glanced up from my article on mitochondrial DNA. A small child was standing in front of me. I didn't look her straight in the eyes because Tanzanian children are usually afraid of wanderers. One of the little girl's hands held tightly on to the skirt of a woman standing next to me, the other she offered to me. I took her hand and shook it. Wasn't she afraid of wanderers? I asked the woman.

"She's not afraid of white, her mother has no color."

"Is she an albino?" I asked.

"Uh, *baba*, father," nodded the woman in agreement, although I'm sure she'd never heard the word "albino" before.

As I chatted with her, the little girl stroked the hairs on my forearm.

An extremely skinny man with sunken cheeks and thin slicked-back hair entered the hall. He was wearing a checked shirt, polyester trousers that were much too short, and sandals. He came toward me, laughing, nodded to the woman next to me, and held out his hand.

"Howell, of the White Fathers," he said.

I stood up and shook his hand.

"I have a flat tire. Do you have a jack?" he asked.

I said goodbye to the woman and the little girl and walked to the Land Rover to fetch my jack.

The priest changed the tire as if he'd been doing it all his life. He had come to pick up some nuns. They were supposed to have arrived on the same flight as my guest. But he wasn't counting on any more flights that day and as soon as he'd finished changing the tire, he drove back into town. I didn't want to give up so easily and wandered around the landing strip. I pulled out my notebook and began writing: "Black, open-billed storks searching for snails in swampy field next to runway ..."

A muscle-bound soldier wearing a loose-fitting khaki suit and black leather shoes without laces walked toward me. In his right hand he carried a rifle. I cursed myself for having written anything. It was the same soldier who, on an earlier occasion, had ordered me to leave my notebook in my bag while at the airport. Nonsense, I had said. Any military power could chart the whole airport, including the molehills under the runway, with satellite photographs, I had told him. And there was nothing more to be said.

"This time you're arrested," said the soldier, almost sadly, as he removed the notebook from my hands. "Have you been arrested before?"

"Of course," I replied, not without some pride.

"So for sure you know the procedure?"

I nodded.

"We'll take your vehicle, it's easier," he said, melodiously.

We walked toward the Land Rover. I held the door open for him, and shut it as soon as he was seated, the rifle clamped tightly between his knees.

"Which way are we going?" I asked, as I started the engine. The soldier pointed straight ahead. I had no idea what was going on a few moments later when he ordered me to drive along a small sandy path. Shaking and bouncing, we drove along the virtually unnavigable course toward the lake. After we had traveled a few hundred meters, he asked me to stop in front of a dilapidated mud hut. He got out and went into the hut. On one of its grayish-brown mud walls were the white letters: "Don't mix business with friendship." Underneath, drawn in blue paint, was a mythical figure, half man, half hyena. Next to it, depicted in charcoal, was a human figure who appeared to have nothing to do with the text or mythical being. The soldier returned after five minutes and got into the car. He had a *shida*, a problem.

"Is it solved?" I asked.

"Yes," he replied curtly. Now to the office that was to handle my case.

I started the engine, turned the Land Rover around, and drove back along the sandy path in the direction from which we had come.

"*Karibu mahindi*, perhaps you'd like some maize," said the soldier, as he unwrapped two roasted pieces of maize from a newspaper. I took one and thanked him.

"From your own field?"

He nodded proudly and asked if I had any salt with me. Unfortunately I didn't keep salt in the vehicle.

"*Simama*, stop," he said in a friendly tone, pointing to the next hut.

Upon hearing the sounds of an engine, a little old woman, as shriveled as a piece of dried passion fruit, appeared in the doorway.

"*Karibu*, welcome, welcome," she said cheerfully. "Do we have guests? Are you coming for tea?"

I sensed that it wasn't for me to speak and let the soldier do it, even though the old woman focused most of her attention on me.

"The wanderers are back," said the soldier, laughingly.

"All the better, all the better," replied the old woman excitedly.

I was confused. Here I was doing my utmost not to be taken for a colonial and an old lady was welcoming me wholeheartedly because the colonials had returned—albeit under arrest. The soldier rattled on about something in a language I didn't understand and then asked for salt in Swahili. A pity, a pity, she didn't have any salt. If only she had known we were coming. She would have bought it. The soldier took leave of the old woman, and then addressed me: "*Twendeni*, let's go."

We continued our journey, eating maize as we went. Out of the corner of my eye I saw the soldier sitting contentedly next to me as if we'd been on holiday together for several weeks. He showed the way to the local police station and I realized we'd already passed it earlier. I parked the Land Rover near the station. At the last moment, just before the soldier took me inside, he asked: "Maybe you have a mattress?"

"A mattress?" I repeated, surprised.

If I would please sell him my mattress before I returned home. Perhaps he could already begin paying it off, otherwise it would be such a large amount all at once.

"I want to make a booking," said the soldier.

I apologized for having to disappoint him. In the first place I wouldn't be leaving for some time, and when I did, I had promised my possessions to the fishermen with whom I was living. They would be angry if I sold my mattress to him.

"Would they be angry?"

The soldier's face clouded over and for a few moments he looked as though he'd been hit, but then it dawned on him.

"Of course they'd be angry. Of course. Say it yourself, wanderer," he ruminated, as we strolled into the office.

The soldier handed me over to a policeman and said farewell: "Wanderer, I'm leaving. *Kazi, kazi*, back to work. By the way, thank you very much."

"It was nothing, it was nothing. *Asante kushukuru*, thank you for thanking me," I replied.

The police chief invited me to take a seat. Looking grave, he leafed through my notebook, which the soldier had given him. Meanwhile I looked around the office. On a small wooden bench against the wall sat three timid-looking men, all dressed in rags. One of them wasn't even wearing shoes, only tattered shorts and a shirt full of holes.

The walls of the office had been covered with plaster and painted a dark yellow. On one of the walls hung a black-and-white photograph of the *mwalimu*, the master, President Nyerere. The wooden chairs on which the policemen were seated were modeled after an old English design. The English had introduced this type of chair during the 1920s, and ever since, the design had been copied as faithfully as possible. Each chair was meant to be a replica of the previous one and so on back to the very first model, which had fluttered down from the British reservoir of ideal objects to settle on Tanganyikan soil.

"Are you dreaming, wanderer? Where do you come from and what are you looking for?" asked the police chief in Swahili, as he continued browsing through the notebook. In one corner of the room lay stacks of files and papers bound together with string. On the old wooden desk was a typewriter and next to it a thin book. Written on the soft cover was: "Help from Above."

"So now my specific question is," he continued in English, "what were you doing at the airport although in fact there was not even an airplane?"

"I was waiting for one," I replied in Swahili.

"But why didn't you wait in the arrivals hall? It was built especially for waiting in. Instead, you went outside and made notes."

The police chief looked at me suspiciously. A second agent, who was now examining my notebook, hit the open page triumphantly with the back of his hand, saying: "Look, here's a drawing of the airport."

He pointed to a sketch I had made several weeks earlier. I burst out laughing and added: "Of course not, that's not the airport. That's the floor plan of my parents' house. That's where I grew up. Look, this is the hallway, this the kitchen, and the piano was in that big room."

"It looks very much like the airport. This is too big ... too big for one house. Eh, eh, eh, is this one house?" said the police chief, surprised. Laughing heartily, he concluded: "We are so backward."

I wouldn't be so sure, I thought, but remained silent. The second officer, who was still mistrustful of the whole business, asked: "Then tell me what this is."

I read aloud: "Third day after fertilization: the eyes are given the finishing touch. Lenses are formed. They can start seeing. If these embryos develop enough vitality, then seeing = feeling."

"That has to do with my work. It's difficult to explain. It's about fish progeny," I said.

"Ah, you're a fish doctor, *mbwana samaki*, our fisherman. Very good, very good. So these MiGs don't have your principal interest?"

"Not in the least," I answered, relieved.

A voluminous agent, dressed in a white blouse and blue-trimmed black trousers that were far too tight, squeezed behind the desk, trying to catch the attention of the police chief. "Later, later, I'm busy with the wanderer right now." She disappeared again meekly into the adjoining room, waving to us with her ample backside as she went.

The first agent, who was rocking on his chair while holding on to the desk with both hands, added cheerfully: "The Russians send us bombers, why do you send us only biscuits? Oh well, it doesn't matter, just tell me how to smoke butterfish. I'm told you wanderers are experts at it."

At that moment a poor handcuffed wretch was pushed hardhandedly into the office by two policemen and taken away through a door. A few moments later the policemen returned to join us.

"Our fish doctor is going to tell us how to smoke butterfish," said the police chief to the two newcomers. "The floor is yours, *mbwana samaki.*"

I proceeded: "Take an oil drum and saw off one end. Make a hole in the side. Then start a fire in the bottom of the drum. Make sure you have a good supply of wood shavings on hand. Cover the top of the drum with a piece of damp burlap. When the wood is still glowing but the flames have dwindled, sprinkle the slightly dampened wood shavings onto the coals until they start to smoke."

"What about the fish?" asked one of the policemen.

The gathering had turned into a regular hen party. With friendly smiles on their faces, the policemen listened to the recipe. The three men who had been sitting on the bench against the wall all this time were listening too. If things continued like this, I would be away from here by nightfall. I stopped talking to listen. Strange sounds that I couldn't place were emanating from the room where the handcuffed man had been confined.

"Is someone crying?" I asked.

"The fish," said the police chief, "tell us about the fish."

"The salted butterfish have been speared onto skewers. As soon as the fire starts smoking heavily, hang the skewers in the drum. Then cover the drum again with the burlap sack."

I now definitely heard moaning and weeping, and asked if someone shouldn't go and take a look.

"Continue," said the police chief, irritated.

"That's it," I replied. "They're done. When all the fat has dripped out of the fish, they're done."

The policemen conversed enthusiastically about the recipe.

"When will you bring us some fish?" asked one of them. "And don't forget the oil drum," added another.

"*Karibu* Tanzania, welcome to Tanzania," said the police chief, as he slammed my notebook shut and returned it to me, "and if you go to the airport again, stay out of the way or wait in the hall. Go in peace."

Blue Café

To celebrate my newly found freedom, I headed for the Blue Café, a small eating place in town. While driving there along the main road, I met our guest, the molecular biologist Verdaasdonk, with his backpack and drum. He had just arrived, and was accompanied by a young man, Jo van Geel, who had a passion for rodents, especially dead ones. In addition to a backpack, which he carried in his left hand, Van Geel was lugging a mail bag from the Rwandan post office over his right shoulder. In it, according to him, were eighty traps.

"I'm sure he'll be able to help you get rid of your mice," said Verdaasdonk. He continued: "An organism only comes to life for Van Geel when it's labeled, soaked in camphor, and lying in his collection."

We drove to the Blue Café, and stowed the visitors' luggage under a small sink in the corner. A minimum of light seeped in through the heavily barred windows hung with sky-blue voile curtains.

We ordered egg rolls, kebabs, and cardamom tea from a skinny African boy with unusually slender hands and long thin fingers. Did the aye-aye, the long-fingered lemur, still exist? Many years earlier I had searched in vain for this already rare animal in the rain forests of Madagascar, where there is a rapidly shrinking radiation of lemurs. One of our distant ancestors might well have resembled these lemurs, which are primitive relatives of the monkeys. It is for this reason that some species are traditionally worshipped as forefathers by the inhabitants of the island. Unfortunately, during my journey I didn't come across a single aye-aye. But the jungle was crawling with biologists in pursuit of them—Danes, Frenchmen, Americans, Dutchmen—all sneaking past each other, ready to embrace the last living lemur.

The long-fingered boy wore a navy-blue overall with sleeves that were much too short and carried a broom made of twigs. Using tongs, a fakir-like Indian with gray hair removed the snacks we had ordered from a glass display case on the counter, placing them on plastic dishes, after which they were gracefully handed to us by the boy. A large poster depicting a group of blond-haired rock musicians in an evergreen forest hung on the grass-green wall opposite us. They were smiling. Our little table with its Formica surface was pushed up against the wall, which was painted light blue and covered with postcards of beach hotels, city maps, and, especially, African animals: marabous, zebras, elephants, cheetahs, leopards, and lions. The rarer the animal, the more postcards of it there were.

Wanderers were meant to save endangered species but instead, endangered species saved one conversation after another between wanderers.

Fearing that the other wanderer would start talking about melting tarmac, souls that allowed themselves to be converted, diesel engines that preheated their fuel, wells that became deeper, or operations carried out by candlelight that ended in bloodbaths, it was always possible to beat the other wanderer to it by asking if he had seen a new animal.

I, too, now put this question to the two visitors who had just completed a journey through several game reserves. Their backs straightened and they began pointing to the postcards of all the animals they had just missed seeing: wild dogs, striped hyenas, leopards.

"But we did see most of them," said Van Geel contentedly, adding: "I know people who were there longer and saw fewer animals."

He removed a pocket guide of African mammals from his bag and began leafing through it. For every specimen of a specific species that he had seen, he had placed a mark in the margin next to the corresponding picture. Humming, he proudly showed the marks. Meanwhile, the molecular biologist explained that he led a different kind of life: one without a guide.

I said I thought that was brave, to which he replied: "Well, I don't think brave is the right word. It's all a question of DNA to me. Everyone needs a straw to clutch on to and mine is a contemporary one. After all, this is the age of reductionism."

"Of rodents," protested Van Geel, without looking up from his guidebook.

Verdaasdonk pushed back his chair, put his hands in his lap, and looked around with a satisfied smile, while Van Geel fussed about an antelope he was sure he had seen, even though it was not marked in the margin. I was just about to ask how long they intended to stay when Verdaasdonk remarked: "In Manyara Park we looked for lions in trees. That's what I wanted to see."

"You're not the only ones," I replied.

The last time a group I was in had been there, our vehicle had been stopped constantly by travelers in other vehicles, who had looked noticeably dour but were otherwise decked out like typical safari-goers, with canvas hats and sunglasses, and stomachs and chests laden with ordinary cameras, video cameras, and telephoto lenses. They had inquired whether we had sighted them—lions in trees.

"No, are there lions in trees here?"

"There certainly are. In this reserve, there's a whole lion population that lives in trees. It's unique."

The fascination with lions in trees turned out to be contagious. From that moment on, the only thing my traveling companions had eyes for were feline tree-dwellers. The atmosphere in the wanderer-packed vehicle became increasingly dejected as the lions proved difficult to find. Scarcely a word was uttered. Finally, someone broke the silence, saying: "It'll be dark in three hours." He continued, after a short pause: "I hope I didn't annoy anyone by saying that." It was silent again after that.

The magnificent landscape shimmering in the heat, the flight of thousands of pink and white flamingos, a herd of African buffaloes that appeared to be casting votes about which direction to move in: nothing made an impression on the travelers from the moment it became known that there were lions in trees. Yet that wasn't the strangest thing. I only discovered what that was late in the afternoon when, to my great relief, I saw a large yellow paw hanging down from a tree. It was about fifty meters from the sandy path. "Stop!" I shouted, and the driver backed up until the paw became visible again. The rest of the lion couldn't be seen from the path because of the dense foliage. My companions, who saw the paw only after I had spelled out a detailed route for their eyes to follow, hastened out of the vehicle, shouting "Wow" and "My God," their cameras poised for action. They then headed straight for the paw, as if nothing on earth would induce the lions to come out of the tree. I had stayed behind on my own and suffered agonies. Why hadn't I kept the paw a secret? To ensure that it would be a bearable evening? How could I ever bring myself to collect the torn bodies and leftover limbs? I looked in the Land Rover and noticed that the driver had taken the keys.

Unscathed and high-spirited, my companions returned to the vehicle some time later, shouting: "You really missed something!"

They'd seen a whole group of lions, including a lioness and her three cubs. After this incident, everybody seemed satisfied and there was endless speculation about the "unforgettable sight."

"It must have been a brilliant lion who first left the ground and offered his kind a totally new perspective," said someone. And wouldn't this culturally transmitted behavior have far-reaching consequences for the lion's physique and way of life? The population might, in the ever-so-useful long term, be able to conquer airspace! Of necessity for example—if the food supply on the ground ran out. Of course, they would have to become somewhat smaller and the skeleton somewhat lighter, and the folds of skin running from the lower jaw to the front legs, from the front legs to the back, and from there to the tail, would have to become wider. The odd muscle here and there in the folds of skin would have to change ... it needn't be too radical a reconstruction ... and there you had it, in no time they'd be gliding through the air. "After all, dinosaurs didn't die out, they flew away," claimed another, with apparent certainty. Singing and joking abounded. Suddenly everyone seemed in a great hurry to return to the hotel. Hours later my traveling companions were still occupied with recording the accumulated evidence, amid contented chatting about "our good fortune."

"That's *maji moto*, the warm water springs in Manyara Park," said Verdaasdonk, pointing to one of the postcards. "Have you ever been there?" he asked. "I put my hand in the water. Seventy degrees Celsius, I think it was. We might be able to find a heat-resistant enzyme there, polymerase."

"Heat-resistant polymerase, what's that used for?" I queried, willingly.

"It's essential for the polymerase chain reaction. Using that reaction we should be able to look at the mitochondrial DNA of the furu to determine whether they are a species flock, a monophyletic group. Or haven't you heard of the polymerase chain reaction yet?"

I had heard about a breakthrough in molecular biology but didn't know any details. While Van Geel, bent over his guidebook, continued grumbling about the impenetrable taxonomy of the yellow-spotted dassies, the molecular biologist informed me of the virtues of the polymerase chain reaction.[31]

The principle of the polymerase chain reaction is simple: it is based on the way DNA replicates in nature. When a cell divides, the double-stranded DNA separates into two single strands. The hydrogen bonds, which in Watson and Crick's model form the rungs of the rope ladder, are relatively weak and are broken. These single strands of DNA act as a template for the complementary strands that, with the help of an enzyme, the DNA polymerase, are transcribed according to the principles of base pairing. According to Zolg, the most striking feature of polymerase is that it can recognize a certain base and that it reacts to this information by joining a complementary base to it. If polymerase recognizes the base guanine, cytosine is brought in; if it recognizes cytosine, guanine is joined to it. Similarly, it brings thymine and adenine together. The result is a double-stranded DNA molecule in each daughter cell, consisting of one strand originating from the parent cell and a newly written complementary strand. The genetic information in the daughter cells is therefore exactly the same as in the parent cell. Before the discovery of the polymerase chain reaction, the quantity of DNA in which researchers were specifically interested was often too small for analysis. The quantity of DNA in the root of a hair possibly originating from a criminal is not enough to allow identification of the person to whom it belongs. The same applies to the amount of DNA found in traces of semen from rapists. The potential applications of a technique whereby a specific fragment of DNA can be replicated are unlimited. It will eventually become possible to identify infections caused by fungi, bacteria, or viruses, such as the human immunodeficiency virus (HIV), which causes AIDS, early on, when the quantity of foreign DNA is still extremely small. Evolutionary biologists will be able to use the technique to replicate a specific piece of DNA, to determine its base sequence, and to compare it with that of other species, to form some idea of the extent to which they are related. And it will become possible to establish molecular phylogenetic trees, in which the bases of the DNA serve as identifying features instead of morphological characteristics. This is part of the reason so many years were spent looking for such a technique. During the mid-1980s, American biochemist Kary Mullis, who received the Nobel prize for his work in this area, stumbled on the desired method. He discovered how, by means of the polymerase chain reaction, a small quantity of DNA could be replicated synthetically into usable quantities. The method works as follows: the separation of the complementary strands of DNA, as initiated by enzymes in cell division under normal circumstances, can be initiated artificially by heating the DNA to 94 degrees Celsius. At this temperature, the hydrogen bonds between the complementary strands are broken, while the strands themselves remain intact.

A specific feature of the enzyme DNA polymerase is that it looks for a place on the DNA—in the form of an irregularity—where it can bind. Only when DNA polymerase has found a small piece of double-stranded DNA, does it begin to transcribe a complementary strand of DNA. It is this property of the enzyme that is used in the chain reaction. Pieces of single-stranded DNA of just the right size are joined to the single strands of DNA—at the beginning of the fragment that is to be replicated in one strand, and at the end of the fragment to be replicated in the other. Despite the hundreds of thousands of bases present in the solution, these pieces of DNA comprising about twenty to thirty nucleotides go directly—and unerringly—to their only goal, namely their own complement. They are called genetic probes or primers. When the genetic probes have arrived at their destination, the DNA polymerase can begin transcribing a complementary strand by pairing bases with each other. Separation of the double DNA strands, marking of the piece of DNA to be replicated by linking genetic probes at the beginning and end of the piece that is of interest, and finally, synthesis of the specific piece of double-stranded DNA—this is the essence of the polymerase chain reaction.

Once the process has been started it can repeat itself dozens of times, whereby the newly formed DNA serves as a template for the synthesis of a new copy of the target DNA. In this way, the quantity of the desired DNA fragment, located between the two genetic probes, is doubled during each cycle.

The polymerase used in the polymerase chain reaction originates from a bacterium (*Thermus aquaticus*) that is found in hot sources, perhaps also in *maji moto*. It can withstand the high temperature necessary to separate double-stranded DNA artificially. By using a heat-resistant enzyme it is possible to produce a hundred million copies of a specific piece of DNA within a few hours.

"A method for tracing a needle in a haystack and subsequently producing another haystack from the same needle, that is how the chain reaction has been characterized," said Verdaasdonk.[31] During the following weeks, he collected an exemplary series of specimens. Unfortunately, he made an error whereby both he and we were temporarily deprived of the benefits of the chain reaction. One evening he forgot to close his drum. The supply of liquid nitrogen shrunk rapidly. In fact, at that moment, everything was still all right because the next morning, as soon as the disaster was discovered, the samples were put into a wanderer's freezer in Mwanza. Verdaasdonk departed and planned to return before long with liquid nitrogen in which to collect his samples. Unfortunately, one day we had to write him and tell him there had been a power failure lasting more than a week and that we feared the worst for his samples. We never saw him again. But it didn't take long before a new generation of bright-eyed and bushy-tailed molecular biologists descended on the furu of Lake Victoria.

Eventually, it was established that the furu of Lake Victoria had evolved from a single ancestral species. It was a genuine species flock.[29] These discoveries dealt the final blow to Greenwood's polyphyletic model. The genetic differences between the furu species were extremely small, even smaller than the differences between certain human populations.

Axel Meyer and his colleagues at the University of California (Berkeley) studied different parts of mitochondrial DNA taken from fourteen species of furu from Lake Victoria. They examined samples from different trophic groups: insect-eaters, fish-eaters, pedophages, algae-scrapers, snail-crushers, and snail-shellers. In total, the base sequence was determined for three fragments—each of which was several hundreds of base pairs long—located in different parts of the mtDNA. Some of the fragments came from the most variable part of the control region, that part of the mtDNA which did not code for protein. Another fragment came from pieces that changed more slowly, such as the so-called cytochrome B gene and two genes that code for transport RNA. The variation found in the base sequences of the fourteen species studied was surprisingly small. Base substitutions were found in only fifteen of the more than eight hundred positions. The variation was smaller than in comparable pieces of mtDNA originating from different human populations, while *Homo sapiens* is in itself a species with very little variation in its mtDNA. But were the hundreds of variations of furu really biological species? Might they not be various biological species that manifest themselves in many different forms? This was unlikely, because, despite the limited variation in mtDNA, no two species were found with exactly the same base sequence. It is virtually impossible for such a result to have arisen by chance. The most plausible explanation is that the different categories of furu identified by us in Lake Victoria were biological species—but very young ones. This impression, already supported by the limited variation in the mtDNA from the control region, was further reinforced by the fact that the mutations found were no more serious than a few base substitutions, while in the more slowly changing parts of the DNA (cytochrome-B gene, tRNA genes) not a single base substitution was observed. Comparisons with the results of other fish studies also suggested that we were dealing with something unique. In general, the variation in the conservative parts of the mtDNA, which was found in different members of one and the same species of mammals, is greater than the variation in the quickly changing control region, which was found in the different species of *Haplochromis*, and even in different genera (*Haplochromis, Macropleurodus, Platytaeniodes*).

In order to classify the furu from Lake Victoria in relation to other cichlids and accumulate additional information relating to the increasingly strong impression that we were dealing with a group descending from a common ancestor, Meyer and his colleagues examined the same pieces of mtDNA in twenty-six other species of cichlids. These species originated from, among other places, Lakes Malawi and Tanganyika and various West African rivers. On average, there was a difference of only three base substitutions between the base sequences of the pieces of mtDNA taken from the furu of Lake Victoria. By contrast, an average of fifty-five substitutions were found between the base sequences of the Victoria cichlids and those of the Malawi cichlids, and at least seventy-seven substitutions between the sequences of the Victoria cichlids and those of cichlids originating from other sources.

From the molecular point of view, the furu of Lake Victoria were closer to the cichlids of Lake Malawi than to those of Lake Tanganyika. They were even further removed from the species from the West African rivers (*Hemichromis*). Species from Lakes Victoria and Malawi bearing a close resemblance to each other in terms of morphological characteristics turned out to differ from each other in molecular terms as much as species from both lakes with very different morphologies. The flock in Lake Victoria displayed a radiation that, although not extreme, was well advanced in terms of morphological features, and showed a considerable differentiation in ecological features. The surprising thing here was not only the explosive speciation and radiation, but the fact that molecular evolution appeared not to have kept pace with morphological evolution.

Geneticist John Avise of the University of Georgia was baffled when he read about this. He responded with, among other things, the following remark: "Morphological and molecular evolution can march to the beat of different drummers."[32] From the molecular point of view, the king crab, a living fossil that has changed little morphologically over tens of millions of years, displays a "normal" pattern of genetic variation in its DNA and proteins, while humans and chimpanzees, which differ so clearly in their physique and way of life, are 99 percent identical in this respect. But even compared with these examples, the divergent morphological and molecular evolution of the cichlids originating from the East African lakes was surprising.

What could be the reason for the mutability of the furu from Lake Victoria? How did the endless series of morphological variations presently found in the species flock develop from a single ancestral form? No one knows the exact answer to this, but the picture was gradually emerging. The following three chapters are devoted to this question, although I will discuss not only the furu but also the theory of evolution in general. Occasionally, the perspective will change too, from that of fish in a closed world, illustrating evolution on a mini-scale, to the general theory of evolution, which clarifies a great deal about these fish. It is difficult to write about evolution in a rigid linear form. This is because the theory is made up of bits and pieces. In my opinion, though, all the circumstantial evidence taken together makes the theory convincing. How does form change, and which mechanism is responsible for changes in form? Is it natural selection? To what extent was Darwin right?

4

The Blister: Evolution through Natural Selection

I sat up carefully in bed to rinse away the bitter taste of quinine, then lay down again and looked outside: grayish-white wisps of clouds against the dark blue night sky. The full moon, scarcely visible, a pale yellow spot high in the firmament.

When I blinked, the spot became elongated, like the reflection of a street lamp on an Amsterdam canal. If I almost closed my eyelids, the vague contours of a fish loomed up. A watery, distorted image that changed shape as more or, in this case, less light reached the retina. Elongated forms, increasingly long, or short truncated forms, ever shorter and more truncated. It was a matter of how tears were distributed over the eyeball.

Impressive, elongated pursuers armed with enormous jaws and teeth passed before my mind's eye. Large hunters lay behind plants in ambush. The most prominent feature of their bodies was their mouths, and they could jump out at any moment to attack. Plump grazers combed algae off rocks, giving the impression that they could continue doing so for millions of years. Snail-crushers with impressive heads shuffled over the sandy bottom in search of prey. The continuous crushing of snail shells produced a deafening, crackling sound. Small, poorly armed mud-biters loomed up occasionally amid particles of mud fluttering to the lake floor. A female with fry in her mouth behaved like a male: aggressive, agitated. She even changed color. Pitch-black, she attacked anything that approached her. Hers was a fictitious territory: it accompanied her wherever she went. She acted like a sexually active male, but was unmasked by a pedophage, a child-eater, who latched onto her mouth and mercilessly aborted her. Within a few days she would no longer resemble a male. Silvery gray, she would be absorbed back into the masses.

Could this enormous variety in structure and behavior be a result of the process of natural selection affecting each of the hundreds of species in this flock? Or was there an alternative hypothesis—a scientific, operational one? Could a diversity of forms as extensive as the one displayed by this species flock also have arisen without natural selection, or was natural selection the most important evolutionary force involved? While questioning the importance of natural selection, I was having the greatest difficulty not falling prey to it myself. Only recently I had narrowly missed stepping on a camouflaged puff adder as it lay in the warm sand catching the last ray of sunlight. If the diversity of forms had nothing to do with natural selection, might it not be related to an as yet unrecognized or unknown form-generating principle? That's what I was thinking when I lengthened my stride to avoid stepping on the completely motionless snake. The world, a minefield of evolutionists? Could I circumvent Darwinism? I would have liked to, but for the time being he kept proving himself right. I saw black.

I turned my head away from the moon and listened to the clicking of footsteps on the concrete. Someone rather slight was walking toward the kitchen.

When I turned my head back, my brain mass seemed to stick to the back of my skull, only to smash against it again a second later. I couldn't bear any more pain. Pain was a warning signal. Fine, now I'd been warned. Let a blister appear. A moment of peace. Or even my final exit, albeit it an untimely one. Blisters were signals, warning signals. Don't go any further, you've reached your limit. Imagine blisters that have developed on your feet. At critical moments, you can ignore them. If you were escaping from a wild predator, the blisters would burst and you would be able to continue running without too much discomfort. An elegant evolutionary solution—with an escape clause: the blister ... I hit my forehead with my fist. Enough of this useless thinking. I couldn't lie down any longer. I sat up, clutched my knees with my arms, and inscribed circles in the air with my head. After a dozen or so circles, the pattern shifted slightly. The only condition was that the line be continuous. Rules were rules, even if I did think them up myself. From the top of the circle inward, to just above the center, and then upward again. A loop, a nose, a humanoid apparition according to a millennia-old, cosmopolitan plan.

Repeat. Keep repeating. It worked. Numbness at last. Now, lie down carefully. The moon was still visible outside. It broke into two segments, then merged, separated, and merged again. Eleven blinks of the eyelid signified a year. I blinked a hundred times to hoard time. Watched myself as I lay in bed. Was amazed by this pointless blinking of eyelids. Water dripped from the melting ice cubes that had been placed on my forehead.

The links were as clear as day. Everything I contemplated was unexpectedly simple. But how to hold on to these insights? How to describe them? As clumsy as the webbed feet that runner ducks have been burdened with—that was the way I experienced the words at my disposal. The external appearance of insight was as deceptive as black ice. I felt disheartened. A vocabulary that at critical moments seemed fit only for doing business or conversing with an insurance agent.

The refrigerator engine went on, making an awful humming noise, but it didn't get any cooler. Steam rose up from my bed as I threw off the sheet, groping for words that would cool me down: sleigh, igloo, cold-blooded, cold storage. Would I be lying frozen the next day, stretched out in the cold storage chamber of the local hospital? Among dead Tanzanians? A power shortage. Bodies crashing against each other in melting ice. Power shortages were the order of the day here.

"Did you say something?" asked a gentle voice, as the door slowly opened.

Sometime later I stood up and left the room. Was that the quiet girl sitting there reading by gaslight? Was she looking at me? Where were her facial features? Something was hanging over her face. She wasn't wearing a veil. Was it the shadow of the window grating? Was she a prisoner perhaps?

"He's confused," said a voice behind me. I turned around but didn't see anyone.

"May I see those footsteps from a moment ago?" I asked the girl. "Do you remember Laetoli? The footprints of the first people, preserved in volcanic ash."

After an eruption of lava, which hardened while still ablaze, the volcano uttered a deep sigh. An ashen rain settled everywhere. It drizzled that day. She didn't listen, but stood up and went to the kitchen.

"The small child who had walked in the footprints of an adult ... Because of that child playing, we may never know how upright they walked then. The first and last footprints in volcanic ash. That they were the same ones ..."

"Be quiet," summoned an impatient voice.

"To give some idea of how short it was ... All in all ..."

"That's enough. Be quiet now."

Shuffling around the room on my knees, I followed her footsteps. Not a trace of light-footedness. I held my head between my hands and rolled onto my back. The concrete felt gratifyingly cool. Again footsteps, first receding, then approaching. She bent over me, bringing her hands to my throat, then smiled and said: "Drink." Despite the noise, the voice sounded familiar.

It irritated me. What I was trying to say appeared not to get through to her: without a sea of time, no evolution. Without time, a beautiful flower would always be the same beautiful flower. Or there would be no flowers at all. No, don't mention all the conditions responsible for the origin of beautiful flowers right now. Don't mention them. Not the wind. The wind produced oak trees and colorless pine forests.

She held her hand to my forehead. It was pounding.

"Easy," she said.

But how to stop it? As soon as I lay down, it started again. Thinking about those who were gentle. Impossible to stop. That the gentle didn't become extinct had to do with the fact that they kept being reborn. Something isn't right in the theory of natural selection. They weren't unusual enough to be mutants. Or could gentleness be hereditary? Or good for something? My eyes saw black again.

"I'm better," I announced. "I'm going outside. It's turning light. Catch butterflies. Action and desire must coincide. If anyone understands that, it's you. Melle's going to the Netherlands. Didn't I say that already? If he would kindly bring some living butterflies for the evolutionary biologists. They are working on natural selection. They use butterflies for experiments on the origin of color and spot patterns on butterfly wings. They study butterflies that mimic other species, as well as the evolution of inconspicuousness, or of its opposite, conspicuousness. Exactly the same things as I am studying in the furu. They are requesting butterflies of the genus *Bicyclus*. They can be found in abundance at the neglected expatriate graveyard."

But she blocked the outside door. She had put the key away in her pocket.

"You're not leaving yet," she said, her voice scarcely audible above the humming. I went to bed again. Had I fallen asleep?

"There, near 1897." I walked around until I was standing behind the stone on which only a date—1897—was engraved. No polished stone, no wooden cross, no name, no mention of *trauernde Hinterbliebenden.** I clapped my hands while Melle brought down his butterfly net a meter away from the stone. He was successful. He removed the orange-brown butterfly from the net and put it into a cardboard box, meanwhile searching for more. "There," he pointed. I walked to the gravestone of August Heimke. Unteroffizier in der Kaiserl. Schutztruppe 1899–1901. Born 29.VI.1875. Died 6.VI.1901. The butterfly had landed in the grass next to the stone ... Brought the net down too soon. Be more careful next time. Melle pointed to the bushes behind Friend Decimus, onto which several butterflies had settled. We stalked toward them: FRIEND DECIMUS (JACK) BRAZIER. Born in Brighton, England: 4 Dec 1908. Died in Misungwi, Mwanza region: 3 Sept 1975. The last British pioneer, prospector, and miner, who spent 40 years in developing gold, mica, and diamond mines in Tanganyika. From 1965 to 1973 he strove with courage and fortitude to work his diamond claims at the Mwamanga pipe near Misungwi, but his efforts were thwarted owing to a change in the country's mining policy. His achievements lay in his finding the new diamond deposit at Mabuki, in 1972, and in helping indigenous people till the end of his life. ... They were too high up, those butterflies. We walked on. No butterfly near Baby Casmiro Francis Vaz's stone, but one did land on the gravestone of a Chinese child. I prodded with a stick in the grass ahead of me as I approached the grave.

"To avoid snakes without missing butterflies. That's the trick," I said.

"Precisely," said Melle. "And that not only applies to you but also to birds that hunt butterflies and don't want to find themselves choking on a snake's head. Do you know those butterflies that, when alighted, look exactly like snake heads? Imitation heads. Can you clap now?"

*German for "bereaved next-of-kin."

I clapped above the butterfly and Melle caught it. His hand disappeared carefully into the net so he could grasp it. He peered at me triumphantly, smiled, and crept forward, toward the Russian, keeping his index finger poised in front of his lips: Eric Novotny 22.3.07–13.1.70, strikes, captures. RIP ... That he may rest in peace. I got the impression Melle could catch butterflies just as well without me but he believed in "collaboration."

Would he get through customs with them? The ticking and flapping in the box was clearly audible. The customs officer put his ear to the box. His face revealed signs of fear, but also of shame: "Spirits." The government had banned all belief in spirits, but it was difficult not to believe in what was present. Omnipresent. Day and night. Proceed, he indicated diplomatically, with a wave of the hand: go, away from this country, into that bird. A more down-to-earth colleague wasn't satisfied and, despite warnings, opened the box. Hundreds of butterflies dispersed throughout the arrivals hall.

The Hospital

When I awoke, I found myself in a short narrow bed, with my left arm attached to a drip and my legs protruding at the end of the bed, like the legs of a chicken I had been served recently in the little No Sweat Café. The thin, elongated legs of the skinny bird had extended at least twenty-five centimeters beyond the edge of the plate. "Is this chicken?" I had asked the waiter, holding my index fingers at the beginning and chopped-off ends of the legs. "It looks more like a heron." The waiter had picked up the menu and read aloud: "Chicken and rice, look, here it is, chicken and rice. That's what you ordered, isn't it?" And, pointing to a pineapple-shaped plastic container, he had added, in deadly earnest: "This is a boiled water."

A small hospital ward. The bed on my left side was vacant. In the bed on my right, an emaciated Tanzanian was tossing and turning under a light-blue sheet. His bloodshot eyes fixed on me with a friendly expression.

"*Umeamka salama*, have you awoken in peace?" he asked.

My mouth was too parched for me to reply. I couldn't separate my jaws and felt too weak to nod, but the pain had disappeared. It was heavenly. On the other side of the ward were more cast-iron beds, also occupied.

A nurse hastened down the corridor. The doors, which opened onto the inner courtyard with its old trees, were wide open. A fresh, gentle breeze carried the intoxicating scent of flowers into the disinfected ward. Apparently there had been a heavy downpour. There were little puddles everywhere. Women were singing and laughing as they did their laundry next to the tap: the sounds of gushing water, of iron buckets being set down, of wet clothing being slapped against a flat rock. How did I get here? "She's here," said the man in the bed next to me, as if he could read my mind.

A few days later the drip was removed and I made little excursions around the ward. A Dutch doctor, a strapping woman wearing plimsolls and a flowery tent dress, appeared at my bedside, after having conversed with my neighbor. She took the time for her patients and made a capable impression.

"You're certainly much better. Malaria tropica. It eats away at you, destroys tissue," she said. "We almost lost our evolutionary biologist. You're working on problems of evolution, aren't you? I've always found it a very primitive concept, the theory of evolution ... Giraffes have long necks because only those specimens with the longest necks can reach the highest leaves. That's what evolutionists claim, isn't it?"

"Darwin wrote about the origin of giraffes to illustrate 'his teachings about descent with gradual modification through natural selection,'" I replied.

> The giraffe, by its lofty stature, much elongated neck, fore-legs, head and tongue, has its whole frame beautifully adapted for browsing on the higher branches of trees. It can thus obtain food beyond the reach of the other ... hoofed animals inhabiting the same country; and this must be a great advantage to it during dearths. ... So under nature with the nascent giraffe, the individuals which were the highest browsers and were able during dearths to reach even an inch or two above the others, will often have been preserved; for they will have roamed over the whole country in search of food. That the individuals of the same species often differ slightly in the relative lengths of all their parts may be seen in many works of natural history, in which careful measurements are given. These slight proportional differences, due to the laws of growth and variation, are not of the slightest use or importance to most species. But it will have been otherwise with

the nascent giraffe, considering its probable habits of life; for those individuals which had some one part or several parts of their bodies rather more elongated than usual, would generally have survived. These will have intercrossed and left offspring, either inheriting the same bodily peculiarities, or with a tendency to vary again in the same manner; whilst the individuals, less favored in the same respects, will have been the most liable to perish.

We here see that there is no need to separate single pairs, as man does, when he methodically improves a breed; natural selection will preserve and thus separate all the superior individuals, allowing them freely to intercross, and will destroy all the inferior individuals. By this process long-continued, which exactly corresponds with what I have called unconscious selection by man, ... it seems to me almost certain that an ordinary hoofed quadruped might be converted into a giraffe.[33]

There is still no better explanation for the giraffe's long neck than this one.[34] In neo-Darwinian terms, his theory boils down to the following: organisms that reproduce sexually nearly always produce more than two, and sometimes even millions of offspring per pair. But despite the production of more than two offspring, densities of species generally fluctuate around a certain natural equilibrium. The reason for this is that many organisms die before they reach the reproductive age. The idea is that they are not just random victims. Through natural selection, those individuals relatively poorly adapted to circumstances are eliminated. If characteristics responsible for the relative adaptedness of the remaining variations can be passed down, the offspring will also have these characteristics; this is the way the gene pool of a population or species changes. As long as the environment does not change, the population will become increasingly adapted to its surroundings.

The doctor beckoned to a male nurse who was passing by with a trolley, and inquired if there were any bandages in the hospital. She apologized to me for the interruption. I followed her eyes to the ersatz purple flowers of the bougainvillea near the washing area. Another case of natural selection. Leaves that acquired a seductive color to lure pollinators. A large woman clad in colorful wraparounds strolled past the bush, humming.

"The principle is surprisingly simple, but a simple principle is not the same as a primitive idea," I concluded triumphantly, believing I had convinced her.

"I'm not impressed," replied the doctor obstinately. "Perhaps giraffes developed long necks because otherwise they wouldn't have been able to reach the scarce water in the ponds where lions were lying in wait. They had to grow long necks to outreach their own long legs, and they developed long legs because only the giraffes with the longest legs were able to shake off the lions. No, it's an entertaining intellectual pastime, that Darwinism, but I have an imagination, too."

Why was this doctor so skeptical? I sensed a certain reluctance to accept natural selection as the most important evolutionary force. It was true that many evolutionary models had been thought up and never tested. Yet, slowly but surely, a respectable number of experimental demonstrations of the importance of natural selection in evolution had been given.[35] For instance, in insects, fish, birds, reptiles, and mammals, including humans.

"The hypothesis of evolution through natural selection is literally too good to be true," the doctor continued. "There must be something else. I simply cannot believe that such a simple mechanism can be responsible for the complexities of nature. How can something so simple produce something as complex as a human being?"

"That sounds like a voice from the nineteenth century. 'Am I standing face to face with someone who still doesn't want to see herself as an ape? Why are you looking at me with that defiant glint in your eyes? You shaggy-haired baboon, who, with your stooped posture, jumps and climbs everywhere with such extraordinary agility ...,'"[36] I said.

But the doctor denied my allegation: "You don't understand. Evolution is a fact. I don't question that. Only the mechanism behind it. That eternal stretching of the giraffe's neck bores me. I want experimental evidence that natural selection is the most important evolutionary force."

I was about to expound on one of the most beautiful examples I knew of—the left- and right-handedness of scale-eaters in Lake Tanganyika—but I hesitated, seeing the expression on her face. Perhaps it would be better to give other classic examples than that of the giraffe's neck: resistance to malaria, mimicry of color and spot patterns on butterfly wings, or deterrent eyespots.

The doctor had to leave, to visit a woman whose skull had been worked over with a *panga,* or machete. Not in a hit-or-miss way but ritualistically, first on one side, then the other.

I let myself fall backward in bed and contemplated the radiation of the furu. The fanning out of these species in countless directions was related to the acquisition and processing of food: mud-biters, algae-scrapers, leaf-choppers, shell-crackers, shell-crushers, plankton-suckers, insect-eaters, prawn-eaters, parasite-eaters, scale-eaters, fish-eaters, and pedophages. A radiation in a closed world, a world in which it was quite inconceivable that such a radiation would not be adaptive. Adaptations that had come about as a result of natural selection. But, admittedly, this was one of the tasks facing biologists: to test, through experimental research, the adaptive value of differences in morphology between species. It was easy to say that morphological differences between the heads of the furu originated through natural selection. It was another thing to prove it.

One of the many hundreds of species of mouthbrooders in Lake Victoria is *H. welcommei.* The individual members of this species have strangely shaped mouths, housing many parallel rows of teeth. Together, these teeth form an efficient rasp. An examination of the stomach contents of several individuals of this species confirmed that it procures its food by rasping. The stomachs were packed with fish scales. This suggests a relationship between certain morphological features of the mouth (the rasp) and their function (the scraping of scales). Perhaps these morphological features (in this case, the specific structure of the rasp) can be explained in terms of function (the demand placed on these fish by the environment, namely, that of being able to scrape scales). But as yet, no relationship has been established, only a correlation between certain morphological features of the mouth and an activity important for the survival of the individual: scraping scales. The adaptive value of the morphological features of the mouth that are related to scale-scraping has not yet been demonstrated.

A hypothetical experiment: a small number of fish tap "scales" as a food source and live exclusively off this source. Although they once ate scales only sporadically, they are now specialized in eating them. What is important here is not their absolute number but their relative number: the number of scale-eaters in relation to the total number of potential victims. Let us assume that, after starting to tap this new food source, the number of scale-eaters increases initially relative to the number of victims, because for the time being scales are available in profusion. But the profusion is limited: at a certain moment there will be a shortage of scales. When scale-eaters have become so specialized that they cannot change to another food source without paying a price for it, competition will arise—competition for scales between members of the same species. It can be expected that efficient scale-eaters—for example, specimens that consume a maximum number of scales within a limited time span—will have an advantage over the less skillful specimens. These efficient and therefore relatively well-nourished scrapers will probably bear more offspring than the less healthy ones. If the ability to scrape efficiently is hereditary, then the scraping efficiency of the population will increase. Given the morphological constraints of the organisms, at a certain point in time it will be impossible to improve scraping efficiency any further. A final evolutionary stage will have been reached in the form of a population of optimally adapted scale-eaters, of eminently suited competitors. There are several examples of species of furu in which foraging behavior has been optimized.[37] But is it really a final stage? Or might it be possible to perfect scraping efficiency even further?

Left- and Right-handedness

Lake Tanganyika is much older than Lake Victoria. This may be why there is only one species of scale-eater in Lake Victoria, while in Lake Tanganyika, there are seven, all belonging to the genus *Perissodus*. The scale-eaters of Lake Tanganyika have a specific feature that is absent in the *H. welcommei* of Lake Victoria: the suspension of the jaws on the cheekbone is asymmetrical, with the jaws of these fish being curved either to the left or right.

Michio Hori spent many years studying the most common scale-eater in the littoral zones of the lake, *P. microlepis*.[38] This species approaches its prey from behind, then attacks it and rasps a number of scales from its flank. Hori conducted underwater experiments in which he offered scale-eaters their prey. He observed that the "right-handed," or rather "right-mouthed," specimens always attacked the left flank of their prey and the "left-handed" ones the right flank. He also examined their stomach contents. The structure of a scale usually makes it possible to identify whether it has originated from the left or right flank of a victim. The stomach contents confirmed his observations of their behavior: right-handed fish had scales from left flanks in their stomachs, and left-handed fish had scales from right flanks.

The asymmetrical structure of the mouth improves the scraper's grasp of its victim's flank, as long as the left-handed scraper attacks the right flank and the right-handed scraper the left. The demand placed by the environment on the symmetrically built *H. welcommei* of Lake Victoria, namely that of being able to scrape scales, was probably once just as prevalent in Lake Tanganyika—until either left- or right-handedness came to be at a premium. There must have been a moment in time when both left- and right-handed individuals had an advantage over nonspecialized scale-eaters. The functional demands imposed by the environment had become more stringent. Young scale-eaters feed on plankton, not scales, during their early days. But even in these young fish, it was possible to establish whether they were left- or right-handed, thus supporting the idea that the type of "handedness" was hereditary.

Crossing of left- and right-handed fish revealed that "handedness" is passed down via a simple Mendelian system (one gene with two possible forms of expression, whereby right-handedness is dominant over left-handedness). Victims were generally alert to scale-eaters. Only a small portion (20 percent) of attacks were successful. This is important. Since "handedness" is hereditary, one would expect there to be as many left-handed as right-handed fish. Why? This is because of a phenomenon known as frequency-dependent natural selection. When right-handed scrapers are in the majority, victims are on their guard for attacks on their left flank. When left-handed scrapers dominate, then prey expect attacks on the right flank. Therefore, during periods of domination by right-handed scrapers, attacks on the right flank by the relatively scarce left-handed scrapers will have a greater chance of success. The idea is that this increased chance of scraping success will have a positive effect on reproductive success, whereby the number of left-handed scrapers will increase. The greater the number of left-handed scrapers, the more alert victims will become to attacks on the right flank. This frequency-dependent mechanism could explain the continued existence of both left- and right-handed fish—an instance of carefully balanced polymorphism.

Hori found that, over a period of eleven years, the ratio between left- and right-handed scrapers hovered around 1:1, never deviating much from this. A scale-scraper leaves a mark on his victim. For a long time the victim retains a scar at the place where the scale has been torn off. During periods when left-handed scrapers were in the majority, the number of scars on left flanks in relation to the number of predators was larger than the number on right flanks. During these periods the victims were less alert to attacks on the left flank. If the right-handed scrapers were in the majority, then the opposite was true. Hori also found indications that right-handed scrapers were more successful reproductively during periods when left-handed fish dominated and vice versa. It is unlikely that there is an intrinsic advantage associated with either left- or right-handedness. This means that the occurrence in equal numbers of left- and right-handed scale-eaters is probably maintained through frequency-dependent natural selection.

This example demonstrates how closely related form and function can be. The difference in the morphology of the mouths between the left- and right-handed varieties can be explained in terms of their different functions, namely those of right- and left-flank scraping carried out by left- and right-handed scrapers respectively. The morphological differences between the different trophic types found in Lake Victoria are probably also the result of natural selection, but no similar phenomenon has yet been studied in as much detail as the "handedness" of the scale-eaters.

"Left-handedness ...," I began, as soon as the doctor came within earshot, but then she was gone again, out of the ward. The gospel of the origin of "handedness" eventually reached her in fragments. She had, after all, asked for it. But was she convinced?

"Even if natural selection has been important in the development of the diversity of form in those fish, this is still very different from something as complicated as human features. You biologists shift back and forth too easily–from animal to human, from handedness to religion, from staying alive to creating art. Human beings are cultural animals. Rapid cultural evolution surely has a much greater impact than endless messing about with genes. But let's keep it simple: give me an example of the influence of natural selection on human evolution."

And off she went again. Before she left the ward, she turned her head in my direction, raised her right hand and said laughingly: "Proof."

I am the best evidence there is, I contemplated. The influence of natural selection on humans has been severely limited by medical intervention, but if I had had a natural defense against malaria, I wouldn't have been lying in this hospital bed.

Evolution is always related to shifts in gene frequencies in populations. Examples of this abound: the increase in the frequency of genes that give giraffes their long necks while genes that would leave them with useless short necks have become rare; the evolution toward a 1:1 ratio of left- and right-handed scale-eaters; the increase, in areas where malaria is prevalent, of a gene that can cause fatal anemia when it is inherited from both parents but that in a single dosage offers partial protection against malaria, and the increasing rarity and ultimate disappearance of that same gene in areas where there is no malaria. Biological evolution always takes place at the level of populations. Was I going to have to tell a doctor of tropical medicine about sickle-cell anemia? Sickle-cell anemia, a fatal disease that, until recently, has plagued doctors of tropical medicine. Of course not. But it is surprising how many doctors still do not realize that the continued existence of the gene for sickle-cell anemia is one of the most powerful examples of the effect of natural selection on human evolution.

Sickle-cell Anemia

Until recently, sufferers of sickle-cell anemia seldom reached the reproductive age. Yet despite this, the gene responsible for this fatal illness has not disappeared rapidly from the population. How is this possible?

As stated earlier, chromosomes are made up of double-stranded DNA. Each strand comprises a sequence of bases, the nucleotides, which, metaphorically, are called genetic letters. A sequence of three genetic letters determines which of the, in total, twenty amino acids will be used for building a certain protein. An example of a protein assembled from many amino acids is hemoglobin, which plays an important role in the transport of oxygen in the blood and a key role in causing sickle-cell anemia. Biochemist Hubert Stryer gave a detailed description of this illness, a summary of which follows.

In 1904, Herrick, a Chicago physician, identified an unusual phenomenon in the blood sample of a black patient. Herrick wrote: "The shape of the red blood cells was irregular. Particularly striking was the prevalence of elongated, sickle-shaped cells." The patient was suffering from acute anemia. Not long after the publication of Herrick's description, more African Americans were found who were suffering from sickle-cell anemia, as the disorder was called. During the course of this century the origin of the disorder has gradually emerged. Sickle-cell anemia is an inherited molecular disease. Patients with the disease have inherited a faulty gene involved in coding for hemoglobin, in a double dose, that is, from both parents (homozygosity). Normal hemoglobin in human red blood cells is called hemoglobin A (HbA). Patients with sickle-cell anemia do not have HbA but a deviant hemoglobin called HbS. HbS only differs from HbA with respect to one amino acid. The specific chemical properties of this one amino acid are responsible for a considerable difference in solubility. When the oxygen level in the blood drops, the little-soluble HbS tends to form a threadlike sediment. This causes the red blood cells to become deformed, giving them their characteristic sickle shape. Propelled by the heart, the red blood cells travel rapidly through the blood vessels. This presents no problem for round blood cells, but the sickle-shaped ones often become lodged in the smaller blood vessels, thus impeding circulation. Sufferers of sickle-cell anemia therefore often end up with damaged organs, such as kidneys. Moreover, sickle cells are vulnerable. They usually have a shorter life span than normal red blood cells.

Offspring who inherit a faulty gene responsible for sickle-cell anemia from one parent and a normal gene from the other parent (heterozygosity) are indeed carriers of the disorder but usually do not suffer from it themselves. Only under exceptional circumstances, such as when climbing a high mountain, might they feel unwell. Only a fraction (1 percent) of the red blood cells of heterozygotes for sickle-cell anemia are sickle-shaped, while that percentage hovers around 50 percent in homozygotes.

Whereas homozygotes only have HbS hemoglobin, heterozygotes have both types (A and S) in approximately equal amounts. This example demonstrates the possible extent of the consequences of having only a single mutation in a single gene.[39]

But it is not my aim to describe the history of the discovery of a molecular illness. What is important here in terms of human evolution is that the frequency of the gene that is ultimately responsible for causing sickle-cell anemia can reach as high as 40 percent of the population in some parts of Africa. How is this possible? A gene responsible for a fatal illness that claims victims at a young age should, through the effect of natural selection, disappear extremely quickly from a population. Only if heterozygotes for sickle-cell anemia produce more offspring on average than homozygotes for sickle-cell anemia and than normal homozygotes can the high frequency of the sickle-cell anemia gene be explained.

The solution was found by Antony Allison. He discovered that people who inherited a sickle-cell gene from one parent but a normal gene from the other (heterozygosity) have a much-reduced chance of contracting malaria tropica, the most life-threatening form of malaria. Malaria tropica is caused by a parasite with an amoebalike appearance (*Plasmodium falciparum*). Malaria parasites multiple in red blood cells, particularly in normal-shaped ones. Sickle cells are seldom chosen because they offer a less favorable environment for the parasite in which to multiply. Powerful natural selection in favor of heterozygotes is the reason for the survival of the sickle-cell anemia gene, a classic example of a carefully balanced form of polymorphism.

In the United States, the frequency of the sickle-cell anemia gene among members of the African-American population is much lower (8 percent) than among Africans in the areas from which African Americans originated. This reduction is the result of the disappearance of the selection pressure that favored heterozygotes for sickle-cell anemia: in most parts of the United States there is no malaria.

Mission Station

A few days later a tawny-skinned old man stood next to my bed. He looked like a windblown rabbi, with his long, snow-white curly hair, gray beard, and stooped posture, and spoke a gibberish that most closely resembled English with a liberal sprinkling of Swahili words. But from his accent I could hear immediately that he was Dutch. Wilfried was his name, Father Wilfried. He lived in the missionary house behind the hospital. Wilfried said how glad he was to have found me. I told him I was Dutch, too, to which he replied: "Well, that's easy, then we can speak Dutch. What's your job?"

"Fish ecology," I said.

"Very good," replied Father Wilfried, "at least people can eat fish." Meanwhile his face clouded over.

"Is something the matter?" I asked.

"Oh, it's not important. I'd heard you were a butterfly specialist. I would have shown you my butterflies." Wilfried pressed his lips together and said: "No problem, *basi*, it's all right."

"I'd love to see them," I said. "Do you have mimetic butterflies? I'm very interested in imitation. They are perfect examples of the effect of natural selection. I myself am studying egg dummies on the fins of furu. Have you ever noticed them? They represent a special form of mimicry."

Wilfried took my arm and said: "Come and see me."

"*Hodi*, anyone home?" I knocked on the wooden door of the low building inhabited by the missionaries. A black-faced monkey with a light-gray coat charged toward me menacingly, until it reached the end of the rope by which it was tethered. This must be Maria, the monkey the missionary had talked about when he had given me directions for finding the house. When Maria had had enough of me and turned her back to me, a light-blue colored scrotum became visible into which two enormous testicles had been tightly packed. I called out "*hodi*" again, this time receiving a reply.

"Come in," bellowed a deep voice.

I opened the door. A stout Tanzanian introduced himself as Father Kibara and said: "*Karibu mbwana*, welcome, sir, please sit down."

"Is Wilfried here?"

"I'm expecting him at any moment," replied the missionary. "Would you like some coffee?" He pushed a thermos toward me and offered me a can of instant coffee.

"Are you British? Oh, I thought you were British. I've been in Britain. In Britain they ask you: 'Would you like tea or coffee?' They want you to take tea, but I like coffee, so I ask for coffee: 'Coffee, if you don't mind.' Then they ask you," and Kibara burst out laughing, "'With sugar or without sugar?'"

He imitated a perfect Oxford accent, while his commentary was in clipped Afro-English: "But this is only the beginning: 'With cream or without cream?'"

By this time Kibara was shaking with laughter: "Mild or hot?"

The fat missionary shook his head from side to side.

"All those questions, *mbwana*, and then, in the end, you get a tiny little cup."

He indicated a minuscule cup between thumb and index finger.

"*Hapana*, no. Those British with their odd customs."

Kibara apologized and stood up. He offered me coffee again and advised me to wait for Wilfried. This was the missionaries' common room. It was where they ate, relaxed, and entertained their guests. It was modestly furnished: a refrigerator, a cooking unit, a tap above a cast-iron washbasin, and, next to the tap, a silver-colored container–the water filter. The dining-room table consisted of four small tables pushed together and covered with a piece of flowered plastic. Wooden chairs hugged the table. At one end of the rectangular room, a "nostalgia" corner had been installed: four hideous armchairs encircled a little Dutch coffee table, on which lay a small lace cloth. Against the wall stood a bookcase filled with one and a half meters of religious texts, stacked between piles of *Het Beste*.* Above the books hung a full-length color photograph of the Pope, flanked on both sides by black-and-white photographs of Tanzanian bishops. The photograph of the Pope hovered above a poster of rippling

*The Dutch-language edition of *The Reader's Digest*.

water, as if the spiritual leader had just arisen from it. Two *tjalks** were sailing full-tilt ahead. *Skûtsjesilen*—Frisian for "sailing races"—was printed at the bottom of the poster.

A slender Tanzanian youth entered the room from a dark corridor carrying a pile of plates in his hand. He began setting the table. He was silent and did his work without looking at me. Only when I greeted him did he nod shyly. "Where is Father Wilfried?" he said, echoing my question. He looked as if he hadn't seen him for years, adding: "Perhaps he's praying, or visiting the butterflies." He tilted his head slightly and closed his eyes: "He's coming." I heard nothing, saw nothing, felt nothing, and then, there he was, standing in front of us.

"Would you like more coffee?" asked Wilfried, as he approached the table. "You wanted to see the diaries?"

"No, I came to see your butterfly collection."

"Oh, yes. That's right."

"Are there diaries?" I asked.

"There certainly are, from the earliest days here. From 1881, when the mission post started. The first diaries were written in French. We stopped writing during the 1950s. There were some complaints from the home front that our tone was too colonial. So we decided to stop keeping diaries altogether. That was easier than adopting a new tone."

"When did you come here?"

"Oh, a long time ago, in 1931. At that time there were still many Tanzanians who had fought with the Germans. They were around when the First World War ended. They were panic-stricken."

Wilfried took the thermos and coffee can and pushed them apart. Nodding toward them he said: "Here was one hill and there was the other. They were on one hill with the German officers and the English and Belgians were on the other. There was a flat area between the hills. Shoot, said the Germans. And the Tanzanians started shooting in all directions."

"Not aiming at anything?" I asked.

"Oh no, they didn't care who they shot: the English, the Belgians; they were all wanderers."

"But weren't there Africans fighting too?"

*Characteristic Frisian sailing vessels.

Wilfried continued, unperturbed: "Then the order came to cease shooting. One of the German officers walked down the hill to the middle of the plain. He was carrying a stick with a white flag on it. At that moment an English officer came down the other hill, also carrying a stick with a white flag. What could this possibly mean, the Tanzanians wondered. But it became even more bizarre. The two officers met each other in the middle of the plain and offered each other a *hand*. Started conversing! More officers went down to join them. There was much handshaking and animated discussion. The Germans asked the English and Belgians for cigarettes and were given them, too. They continued talking as they smoked, and then the message came: 'The war is over.' The Germans had surrendered. The Tanzanians suffered agonies: would they be slaughtered or enslaved? Of course not. There was a party. Food and drink were brought in. The Africans, who had fought for the Germans, English, or Belgians, danced all night. What did it matter who was to become their next ruler? ... Shall we visit the butterflies now? Perhaps you're curious if there are any hopeful monsters in my collection?"

Father Wilfried looked at me inquisitively and stood up.

I followed suit, replying: *Natura non facit saltum* ("nature takes no leaps"), to show that, with all respect, I did not share the evolutionary ideas of my namesake.

Wilfried frowned and said: "A prominent name in the butterfly world, Goldschmidt."

I sighed. The hopeful monsters had pursued me even into the heart of Africa.

Geneticist Richard Goldschmidt worked with the lepidoptera (butterflies and moths), especially with the moth *Lymantria dispar.*[40] In 1940 he published the book *The Material Basis of Evolution*, in which he explained his unorthodox view of evolution.* What did this contrary figure maintain who would not allow himself to be defeated by an almost total absence of followers? Goldschmidt believed he had found evidence for the occurrence of mutations that had a much greater effect than the neo-Darwinists were even prepared to consider possible for a single mutation. He believed that such macromutations—and not a collection of favorable but small mutations accumulating over a long period of time—were the driving force behind fundamental anatomical or physiological reorganizations of organisms. Every so often, the hybridization of two average parents could result,

*See *Time Frames* by Niles Eldredge, which gives a fuller treatment of R. Goldschmidt's theories. The factual information about R. Goldschmidt was taken largely from this book.

totally unexpectedly, in a new form. These macromutants, the so-called hopeful monsters, were often not viable enough to survive or led a life of ill health, but occasionally an individual appeared that seemed cut out for life. A new form or even species—suddenly, from one generation to the next.

Goldschmidt's macromutations greatly upset the neo-Darwinists. However common these mutations may have been in Goldschmidt's mind, they could not be found anywhere else, maintained the neo-Darwinists. And even if such a promising monster did exist, with what could it crossbreed to guarantee its genetic continuity? Just at the time Goldschmidt published his volume, Mayr and Dobzhansky wrote books in which they integrated the Darwinian theory of evolution with genetics—Mendel's Laws and the mutation theory. While formulating this new synthesis, these neo-Darwinists also went all-out to completely discredit Goldschmidt's body of thought. After extensive research, Goldschmidt had come to the conclusion that the mechanisms responsible for genetic variation within a species were not the same as those responsible for genetic variation between species. The neo-Darwinists wouldn't hear of it. Yet whatever intrinsic value Goldschmidt's work may have had, it certainly had at least one virtue: in formulating how fundamentally their opinions differed from those of Goldschmidt, the early neo-Darwinists discovered how much they totally agreed with themselves.

The Butterfly Cabinet

We had entered a small, dark corridor with a low ceiling, just high enough so I didn't bang my head. The air was heavy with the nauseating odor of pesticides. On either side of the corridor were the missionaries' cells. Father Wilfried stood outside the last cell on the north side and searched in his shabby gray trousers for the keys to the butterfly cabinet. The silence was oppressive. He raised his finger to his lips and excused himself, whispering: "The key to the cabinet is still on the table in the common room." Before disappearing down the dark passageway, he glanced hastily over his shoulder, adding: "Back in a minute."

On a piece of cardboard fastened to the wall with drawing pins was the missionaries' daily schedule. It had been handwritten in a rigid script, with the letters adhering faithfully to the predrawn pencil lines:

5:45 rise in silence
6:15 liturgical morning prayer
6:30 meditation
7:00 eucharistic celebration in chapel
7:30 breakfast
...

The missionary returned, saying: “I often get up an hour earlier so I can visit the butterflies.” Then he continued: “I follow Mendel’s example. Did you know he too lived in a religious community? If I hadn’t become so attached to this place, I would have visited his cloister in Czechoslovakia. The garden where he conducted his experiments with peas still exists.”

Wilfried rattled on as if, after years of forced silence, he finally felt at ease to talk about the ungodly subject that held his true interest. “Did I tell you that I started as a biologist? I was a teacher. That was a long time ago, though. Now I’m only allowed to do the hospital administration. Orders from Rome.” He came and stood so close to me that I couldn’t bear it and took a step backward. He didn’t keep the critical distance typical of wanderers—in this respect he had become Africanized. It struck me that I wasn’t bothered when Africans stood so close to me but I was when this Africanized missionary did. Meanwhile, Wilfried continued in a whisper: “I have no difficulty with the theory of evolution. I feel like a wolf in sheep’s clothing here. How do you say that in Dutch?”

We entered the room. It was dimly lit. The atmosphere was heavy with the smell of camphor. I felt dizzy and quelled the urge to vomit. The revolting taste of quinine again. I breathed deeply through my mouth in order to smell as little as possible. There was scarcely any room for standing. Against the walls and in front of the window were cupboards with flat drawers. In the middle of the room was a motorcycle, an old BMW, mounted on wooden blocks. The wheels had been removed and lay on the ground. Wilfried’s voice droned on behind me: “To protect living butterflies from death, and dead ones from life. How do you manage that with your fish?”

Wilfried struck a match. The glowing sulfur head broke off and flew through the air, skimming my face. The next two matches refused to ignite, but the fourth one burst into flame. Wilfried absentmindedly handed me his matches and lit a gas lamp. There was a picture of an elephant on the matchbox and under it the Swahili word *tembo*. *Säkerhets tändstickor*, the Swedish safety matches we used at home—if only I had them with me now. Only a few more years and, God willing, I would be holding the little matchbox with the swallow insignia in my hands once again.

Wilfried closed the door and opened one of the hundreds of drawers. Speaking somewhat louder now, he said: “It happened quickly, didn’t it?" I looked at him questioningly. "At the end of the eighteenth century, a treatise was published on the subject of which language God spoke to the first people in paradise. At the end of the eighteenth century! That seems like yesterday to me. Even when I arrived here in 1931, there were missionaries who still believed that all species had been created by God in that one week. That all species were equally old and had not changed since creation, six thousand years before."

Wilfried stood on his toes and placed the gas lamp on top of one of the cupboards: "Idealism in the Platonic sense was not uncommon in our circles. Individual variations within a specific species were considered imperfect earthly replicas of the ideal individual of that species. What? *Basi*, enough." He stretched, then stood there like a general, back arched, hands on hips, in front of his mounted butterflies. There was no stopping him now. He hurtled on: "In other words, individual variation was deprived of any significance ..." The missionary pointed to a drawer housing dozens of butterflies of the same species and slowly ran his index finger zigzag along the rows. "But, of course, I don't need to tell you anything. You're an expert."

"Quite," I said. "Darwin put an end to thinking in terms of molds. It was a major accomplishment."

Wilfried searched through a box of index cards. I had the impression he wasn't listening. "I want to show you something," he said, "but storage and retrieval were never my forte. Can you stay a while longer?" The missionary flipped eagerly through the cards, stopping every now and then to lick his right index finger. Then he resumed talking: "Imitation, new cases of mimicry. That's what I'm looking for. Batesian imitators. That interests you too, doesn't it?"

"Certainly," I replied. "I'm interested in evolutionary principles, not just fish."

It was silent for a few moments.

"So, if I understand you correctly, the wolf in sheep's clothing is looking for the sheep in wolf's clothing?" I cited my reference: "Rothschild 1972. Are you familiar with that article?"[41]

Wilfried looked at me, puzzled, then pensively, and, finally, with an expression radiating both incredulity and hope: "*Papilio dardanus?*" he asked, cautiously.

"*Papilio dardanus dardanus,*" I replied, without hesitation.

"*Polytrophus,*" continued Wilfried, impassioned.

"*Tibullus,*" I said, doing my best not to flinch a muscle.

"*Cenea?*" said Wilfried, looking at me almost beseechingly as he uttered the name.

I cleared my throat and finally spoke the magic word: "*Meseres.*"

"*Papilio dardanus meseres.* It's here," said Wilfried. Mumbling to himself, the missionary clasped his hands and, lacking space in which to move, rocked from side to side. He touched my left forearm with his fingertips, then quickly pulled back his hand, murmuring: "*Basi*, enough." Like someone possessed, he began opening one drawer after the other.

Imitation and Warning

Wherever I looked in East Africa—on land, in the air, underwater—I was forever confronted with the importance of natural selection. I was becoming increasingly impressed by the Darwinian theory of evolution. Everywhere I looked, I saw traces of natural selection: furu mouths perfectly designed to allow for eating a specific kind of food, Africans much better equipped to resist malaria than I was, mimetic butterflies and deterrent eyespots on butterfly wings hidden away in a cupboard drawer.

Mimicry that offers protection is still seen as one of the most convincing examples of the importance of natural selection.[42] Three years after the appearance of Darwin's *On the Origin of Species*, Bates published his mimicry theory, explaining the extremely strong resemblance between a number of unrelated butterfly species from the Amazon area.

There are three players in the Batesian system: the *model* itself, a poisonous or inedible butterfly species with conspicuous color or spot patterns; the model's *mimic*, an edible butterfly species; and the *predator*, in this case, insect-eating birds. It is in the interests of the mimic to resemble the poisonous variety as closely as possible in terms of external features, such as color and spot patterns, so that the predator cannot distinguish between the mimic and the model. The poisonous model, on the other hand, is best served if the predator *can* make the distinction. These represent the ingredients of a coevolutionary race.[43] Metaphorically, a poisonous butterfly species tries to shake off an edible mimic by altering its appearance every time the mimic begins to resemble it so closely that the predator can no longer tell them apart. In her book *Hedendaags fetisjisme* (Contemporary fetishism, 1925), about man's desire for social distinction, Dutch author Carry van Bruggen wrote: "The girl wears her watch on her wrist, the woman pins it to her chest; the girl pins it to her chest, the woman hangs it around her neck; the girl hangs it around her neck, the woman wears it on her wrist."[44] Although mimicry is not aimed at acquiring and maintaining social status, the interests of the woman and the poisonous model (in fact, no more than a means of inducing vomiting in insect-eating birds) are to some extent similar: both of them want to be noticed and not mistaken for anybody else.

The predator plays a key role in the Batesian system.[45] If there were no insect-eating birds or other predators involved, there would be no point in the imitator mimicking the poisonous model, and no similarities in appearance would arise, except, occasionally, by chance. The predator is essential for the evolution of similarities in appearance between the poisonous sample and the mimic. The bird that is searching for edible butterflies makes a choice. It chooses—from the available prey—those that will be eaten and those that will not. It compares and judges. The bird personifies a selection pressure.

Butterflies are inedible when, during the caterpillar stage, they inhabit poisonous plants, eat them, and store the poisonous substances in their bodies, without succumbing to the poison themselves. These poisonous butterflies are often conspicuously colored. Birds learn to avoid them at an early age. Mimics are butterfly species that, during the caterpillar phase, do not live on poisonous plants, but appear poisonous. They are opportunists.

If poisonous butterflies occur in a certain region where mimics are absent, birds quickly learn to avoid this poisonous prey. The number of victims that die at the hands of insect-eating birds is extremely limited. The greater the number of edible mimics, however, the more diffuse the learning process of the predator. The birds are regularly misled. They become confused and kill many butterflies before beginning to associate color and spot patterns on the wings of the inedible butterflies with a noxious taste. In this way successful Batesian mimics put not only the poisonous butterflies they are imitating at a disadvantage but also the insect-eating birds, in the sense that the latter mistakenly avoid potential prey.

Even until recently, the opponents of Bates's mimicry theory have continued to refute the idea that the origin of similarities in appearance between poisonous and edible butterfly species must be attributed to natural selection. In their view, the selection pressure exerted by birds on butterflies is not strong enough to produce such an effect.[46] But this criticism does not hold water. Monarch butterflies hibernate in dense colonies. The butterflies lie so close together that they form a blanket of color. It has been established that in this type of colony, comprising several millions of butterflies, insect-eating species of birds kill about 9 percent of the hibernating butterflies—proof that birds, in principle, can exercise a strong predation pressure on insect populations.

The error that can become fatal to a predator is that of never making an error. Making errors is perhaps the only way to discover things. Mimicry researcher Huheey, who studied the importance of making errors, pointed out the following: by occasionally attacking a butterfly that it expects to be poisonous, an insect-eating bird will discover shifts in the relative frequency of poisonous and edible butterflies. Insect-eating birds and other predators may even be genetically programmed to make errors occasionally. If there is a large increase in the number of edible butterflies relative to the number of poisonous ones, there is a good chance the young birds will never learn to associate the imitative color and spot patterns with poisonousness. Moreover, adult birds that make errors will become progressively less able to make the association. The result will be that the birds will no longer avoid any prey. The amount of time required before a bird forgets its experience with the poisonous type of prey depends on the quality of its memory, but also on the distastefulness of the poisonous butterfly and the relative frequencies of the edible and poisonous butterflies.

Batesian mimicry is thus frequency dependent. If the number of edible butterflies increases disproportionately in relation to the number of inedible specimens, the mimics will no longer benefit from the protection derived from their resemblance to the poisonous variety. Edible butterflies that imitate a poisonous species must not flourish too readily or they will suffer. Is this always the case? No, not always. There is an escape clause that allows for a sizable increase in the number of mimics.

Batesian mimics—edible butterflies—sometimes represent a problem for more than just one poisonous species. By imitating several poisonous species, the number of edible mimics can increase more than if only one poisonous species is imitated, while, at the same time, retaining an effective mimicry system. The burden of resemblance is shared. A classic example of this is *Papilio dardanus*. The color and spot patterns of *Papilio* females differ from region to region: there are at least eight varieties and each variety has its own characteristic color and spot patterns on the wings, corresponding with the patterns on the wings of a poisonous species.[47]

There are also mimicry systems based on the inedibility or poisonousness of all the participating species. Advertising poisonousness unequivocally and thus easing the learning process of the predator can be to the advantage of several poisonous species. They join ranks under a single "flag," symbolizing the message "poison," so that the predator will generalize. This form of convergent evolution, whereby color and spot patterns of unrelated species increasingly resemble each other, was first discovered by Müller and was named after him. Even an overall resemblance between two poisonous species can lead to convergent evolution in terms of external features.

In Müller's mimicry system, unlike Bates's, the predators (such as insect-eating birds) are not deceived but duly warned. Nor is Müller's mimicry frequency dependent: the effectiveness of the system does not diminish when one of the participating species increases in number. If the resemblance between different butterfly species is unmistakable yet far from perfect, it is likely the butterflies belong to a Müllerian mimicry system. Müller's mimicry stimulates the evolution of uniformity in color and spot patterns, but the mimesis attained is less perfect than that in Batesian mimicry. The predator's capacity to generalize is "anticipated," while the Batesian imitator has to "compete with" the predator's ability to make distinctions. Only in the latter case does the evolution of exact mimicry result.

How does a poisonous, conspicuous species evolve from an edible, inconspicuous or perhaps even camouflaged species? What comes first: poisonousness or conspicuousness?[48] Broadly speaking, there are three models. In the first model—the classic one—conspicuousness and poisonousness evolve simultaneously. But not long ago it was acknowledged that this hypothesis on the evolution of conspicuousness is not the only conceivable one. Conspicuous coloration can also occur without the (in)edibility of the prey being involved. For instance, the demands of sexual attractiveness or the regulation of body temperature (black to absorb warmth) can stimulate the development of conspicuous color and spot patterns. According to this hypothesis, the poisonousness of conspicuous prey only arises later. Predators can learn to avoid conspicuously colored poisonous prey. This is known to be the case in birds, fish, reptiles, and mammals. But the fact that a predator can learn such a thing does not prove that the conspicuous colors of the poisonous prey were developed specifically for the purpose of advertising poisonousness.

According to the third model, conspicuousness only evolves after poisonousness has developed. Because it is poisonous, the organism can risk being conspicuous. Conspicuousness backed by poisonousness.

Eyespots

African butterflies of the genus *Bicyclus*[49] occur in areas with alternating wet and dry seasons. They are common in the Mwanza region, and were particularly prevalent near the neglected wanderers' graveyard. Subsequent generations of these butterflies are confronted with radically different living conditions: cool, wet, and green, or hot, dry, and brown. Some butterfly species escape the long dry periods by flying to wetter climes; others revert to dormancy. *Bicyclus anynana* does not escape the drying vegetation either in terms of time or space. Instead, its external features change radically. The underside of the wings of butterflies that emerge during the dry season has a uniform brown color. These butterflies are scarcely visible among the dead brown leaves—the environment in which they lie dormant, wings folded, for most of the dry season. This cryptic appearance, or phenotype, contrasts sharply with the conspicuous color and spot patterns characteristic of the generation that emerges during the rainy season. These butterflies have conspicuous eyespots comprising a white center surrounded by two concentric circles—a black inner one and a gold-colored outer one. They are more active during the rainy season than during the dry season and rest for short intervals on plants. As long as they do not stay on one plant for too long, the eyespots probably have a deterrent effect on insect-eaters.

a. Rainy-season variety.

b. Dry-season variety.

Figure 4.1
Phenotypical plasticity of the Malawian butterfly (*Bicyclus anynana)*. The phenotype of successive generations differs greatly between the rainy season (above) and the dry season (below). Photo courtesy of Biology Department, Rijksuniversiteit Leiden.

For some time it was assumed that the conspicuous wet-season appearance would not be optimally functional during the dry season, while the cryptic appearance of the dry-season variety would be noncryptic during the wet season (and without the deterrent effect).

This hypothesis was tested a few years ago in Malawi. The field equipment consisted of a butterfly net and two felt-tipped pens. One pen was filled with black ink, the other with brown, a color that blended with the brown on the wings of the dry-season variety. After hundreds of butterflies of the dry-season type were caught, the wings of a number of specimens were marked with conspicuous black eye-spots. The remainder were marked with brown spots as a form of control. The butterflies were then released. Those that were recaptured were examined to determine which portions of the brown- and black-marked varieties had disappeared. The black-marked variety had declined most, furnishing a strong indication that the dry-season species was indeed cryptically colored and that this crypsis was a product of natural selection.

In an environment in which appearance changes radically according to season, there is a powerful selection pressure to produce appearances (phenotypes) well suited to the season. The ability of one and the same genotype to produce different phenotypes as a reaction to changes in the environment is known as "phenotypical plasticity."

The temperature of the environment to which the organism is exposed during its early development is the main factor in determining which appearance the butterfly (*Bicyclus anynana)* will assume. Normally speaking, camouflage and warning colors are two extreme results of natural selection, but, depending on the season, successive generations of these Malawian butterflies are capable of adopting either one or the other of these appearances.

The counterpart of the eye-spot, with its deterrent effect, is the egg-spot. The egg-spot is every bit as attractive as the eye-spot is off-putting.

No, This Is Not an Egg

Whoever had a boat could, without danger of contracting bilharzia, swim near one of the uninhabited rocky islands in Mwanza Gulf. There were no bilharzia parasites there.

Figure 4.2
Egg-spots on the anal fin of a furu male. The egg-dummies of most species closely resemble the eggs of the furu females, at least as perceived by the human eye.

At one time I had intended to study furu behavior either with a snorkel or by diving, but visibility was too limited for my purposes. Whoever lay absolutely still in the water would observe fish swimming by once in a while, but usually they shot away forever into the murky water the moment they were taken by surprise. Something *was* visible, though, something very unusual: fluorescent yellow spots shooting like arrows through the water, then curving out of sight, coming into view again, hanging motionless, then shooting off again. The stiller I lay, the more spots loomed up out of the dark water. I felt watched, surrounded by fish whose only part I could see were the yellow spots, the egg-spots. Egg-spots give the illusion of reality. They belong to the domain of imitation or mimesis and are located on the anal fin of the males. What are these depictions of eggs doing there?

In advanced mouthbrooders, such as the furu of Lake Victoria, pair-bonding has disappeared. Females prefer not to be in the company of sexually active males. They only visit them briefly in order to get milt, staying no longer than is necessary.

The male defends the area surrounding a nest pit, which he keeps clean. With his courtship dance, he attempts to lure female members of his species who are carrying ripe eggs. To this end, as soon as he sees a female, he hovers in front of her with his head turned away and his anal fin, spread and quivering, facing her, so that the egg dummies depicted on it are clearly displayed, his message being "follow me." This is an important function of the egg-spots: to attract females. Sometimes the male shuttles back and forth several times between his nest pit and the female. If he is too slow, he runs the risk of losing her to a rival male. Although the projection of human characteristics onto animals is taboo in ethology, the temptation to do so is great. I once saw a female wait—yawning repeatedly—for a male who had been absent for some time. Slowly but surely she turned away from him, eventually taking off with a fierce, leading neighbor, who took her to his nest pit. Roughly speaking, what happens is this: the female circles above the nest pit, followed in close pursuit by the male, who gently bangs against her genital papilla. If the female is ready to spawn, at a certain point she will lay several eggs, after which she stops swimming in circles, turns around, and takes the eggs into her mouth. It is essential she collect them quickly, because there is an imminent danger of losing them. Egg thieves often lie in ambush.[50]

The quicker the female collects the eggs, the smaller the chance of losing them, but in doing so she also creates a new problem: the male has not yet had the opportunity to fertilize them.[51] During the evolution of mouthbrooding, the moment when the female starts taking the eggs into her mouth has been steadily moved up in courtship. The females of what are assumed to be the primitive mouthbrooders—usually riverine fish that have retained some traces of pair-bonding—allow the males the opportunity of fertilizing the eggs before they take them into their mouths. But in advanced mouthbrooders without pair-bonding, the eggs are collected by the female as quickly as possible, often before the male has been able to deposit his sperm. The males of these mouthbrooders do not participate at all in protecting the fry or caring for the young. The sole extent of their contribution is to pass their genes on to their offspring. But how does the sperm enter the female's mouth? How can a situation be avoided in which females end up swimming around fasting while their mouths are overflowing with unfertilized eggs?

Courtship is interrupted when the female takes the eggs into her mouth. The male now holds his expanded anal fin just above the substrate, beneath the nose of the egg-seeking female, and again begins to swim in circles above the nest pit: he reactivates the courtship by arousing the female's interest with one or more depictions of precisely the thing she is most obsessed with at that moment—eggs. This may well be a function of the egg-spots: to re-establish contact with the female after an interruption in the courtship, to prevent her from laying another batch of eggs in a neighbor's nest pit, though this often happens anyway. After the female has collected the real eggs in her mouth, she reacts to the egg-spots as if they were eggs, trying, with concentration and often, devotion, to bite the dummy-spots on the fin with her lips. But if real eggs were present, she would never make this error; it is as if she is familiar with Magritte's paintings and knows: no, this is not an egg. At that moment in the courtship, Wickler imagined, the male releases the sperm, which then enter the female's mouth and fertilize the unfertilized eggs. In at least one mouthbrooding species, *Pseudocrenilabris multicolor,* it has been established that about half the eggs are fertilized after the female has taken them into her mouth.[52] This can be seen as a confirmation of Wickler's idea that another function of egg-spots may be to increase the chances of fertilization. Yet this is not certain. If the egg-spots are removed from the anal fin of the male, fertilization of the eggs takes place anyway.[53] At least this has been shown to be the case in *Haplochromis elegans,* the species used in these experiments.

Egg-spots are found only in mouthbrooders. All the furu of Lake Victoria are mouthbrooders and have one or more egg-spots on their anal fins. The spots (orange or yellow) are shaped like real eggs and also resemble them roughly in color (yellow or light-brown). The spots on the males are conspicuous, sharply outlined and encircled by a ring in a contrasting color; the spots on the females are much vaguer owing to the absence of a ring and colors. The egg-spots on the males resemble true-to-life replicas of the real eggs, at least to the human eye. How did these tiny naturalistic paintings come into being?

The quivering display of the spread anal fin also occurs during courtship in other species without egg-spots. The anal fin of such species, and probably also of the forefathers of those species that presently have egg-spots, is covered with small, pearl-shaped spots. Females eager to take eggs into their mouths might mistake one or more of these pearl-shaped spots on a particular male for an egg. The idea is that, by chance, spots on some males more closely resemble real eggs in terms of size, shape, and color than spots on other males. If, because of this, the males are more likely to be chosen by the females, and their sons inherit spots bearing a greater-than-average resemblance to eggs, the evolution of the egg-spot will have begun. But is this an example of sexual selection? If so, is it not misplaced in a chapter on natural selection? And is sex not natural? It is difficult to imagine anything more natural. This is true, but on the basis of sound arguments, Darwin (1871) elevated sexual selection to a category of its own within natural selection, a matter that will be discussed subsequently in more detail.

Malaria parasites in my blood; the giraffe's neck; left- and right-handedness in fish; a gene for a fatal illness—sickle-cell anemia—that persists in areas where malaria is prevalent; edible butterflies that imitate poisonous models and are therefore protected from birds; cryptic color patterns on butterfly wings during the dry season but warning colors in the rainy season when this offers better protection against predators ... I was constantly being confronted with the overwhelming importance of natural selection. But did natural selection explain the radiation of the furu? To a certain extent, it did. The fanning out of all these species was probably a consequence of natural selection, as was the differentiation of left- and right-handed scale-eaters in Lake Tanganyika. But the fanning out of species is not the same as the origin of species. Of course, individuals of a single species in two geographically separate areas can become locally adapted if they are exposed to different selection pressures. In this way, natural selection could lead to a different development in the mouths of snail-crushers in two populations of the same species. This would be the case, for example, if one population lived in an area without snails, forcing it to concentrate on another food source, and the other had to survive solely on snails. The divergence in the morphology of the mouth could eventually lead to the development of two new species. In this case the development of a species goes hand in hand with the development of adaptations to diverse conditions. These were the lines along which Darwin thought. This is the way it can happen, but it can also happen very differently. What would have happened, for example, if Darwin had gone not to the Galápagos Islands but to the East African Great Lakes? Perhaps he would have stumbled on the role played by sexual selection in the origin of species. Perhaps he would have discovered that females, by being fussy in their choice of a mate, could initiate the origin of species. Perhaps we would have a different theory of evolution today.[54]

5

A Kiss on the Hand: The Origin of Species

There was a telegram from Leiden in the mailbox: "Maximum width/ height of boat? Forgot. Urgent."

My work was forever being interrupted. They had come up with another idea. The spots could wait; now we were going to do something really important. The project was to be expanded. The seaworthy vessel was more or less ready. Unfortunately, no one had bothered to figure out whether the boat, to be shipped by train from Dar-es-Salaam to Mwanza, would fit through the smallest tunnel. Oh well, not to complain. Not everyone had to share my predilection for the useless, and besides, there was something to be said for the boat. I couldn't go beyond Mwanza Gulf with our little wooden vessel, but once the trawler arrived, all the spots would be within reach. I would be able to probe around in the remotest corners of the lake, at any depth I chose. So, off I went to find the measurements of the narrowest tunnel.

Melle and I cycled to a branch office of the railway station located on the shore of the lake. Before going in, we greeted a guard who was keeping an eye on the long, grayish torpedo boats moored at the wharf.

Inside, in a dimly lit space resembling a cross between a shed and an office, were three hardwood desks with writing surfaces of dark green leather. The racks along the walls were filled with gray files. Place-names were written on the spines: Dodoma, Tabora, Kigoma ... Two of the three desks were unoccupied; a young woman with short curly hair was working at the third. She looked up briefly with watery, bulging eyes as though she'd just emerged from the deep. She was drawing margins in a notebook. As soon as one line was finished, she would turn the page and, in an agonizingly slow tempo, begin drawing another on the reverse side. Melle and I went and stood in front of her desk. I waited, holding my breath. Halfway through a line, she put down her pen. Her lips parted but closed again without uttering a word, only to hang half open again a moment later.

Speech? A relic of the past. Finally something began to vibrate deep in the throat of this moray: "*Shida gani*, what's the problem?"

"Excuse me for interrupting you. We've come for the measurements of the narrowest tunnel between Dar-es-Salaam and Mwanza. It's urgent."

The woman giggled as if I'd said something obscene: "You want its measurements?" Nodding toward a door, she said: "*Huyu*, you'll have to see him. He's the expert."

I was ashamed of my scanty knowledge of Swahili, but "shimo" for tunnel turned out to be quite accurate. The woman shuffled so relaxedly toward the door that I was surprised she didn't collapse in a heap before reaching it. She listened at the door briefly, then knocked and opened it. Melle and I entered. Behind the desk sat a civil servant with a startled expression on his face, as if it was the first time he had seen his door open in years. A warm wind rustled the piles of papers on his desk. There was a window with no glass in it. Thieves could enter easily through it, but there was nothing to steal except a desk, two chairs, and some files. In the Netherlands, thieves broke glass in search of other things. Here, it was the glass itself they were after. The pane had been carefully removed from its frame.

"Ah, it's you," said the civil servant, whose timeworn face was held aloft by a youthful body. He took Melle's right hand and eagerly started pumping it. He knew Melle from town. "Fertilizer? Wells? Cotton? How can I help you?" he asked.

"The measurements of the narrowest tunnel between Dar-es-Salaam and Mwanza. That's what we need."

The man took us to see a more senior colleague who not only had his own office with a desk but a telephone as well. He was small and had a slight build. His bullet-shaped head, of a blue-black complexion, shone like a wet olive. His suspicious glance darted nervously between Melle and me. We were offered a seat and, for the third time, I explained what we had come for.

He interrupted before I could finish: "It's impossible to give you this information now. We need headquarters. If you could come back in two weeks ..."

It sounded like a salvo of bullets from a machine gun. There was no way we could wait for two weeks. Could he perhaps phone headquarters in Dar-es-Salaam, or, if need be, write a letter? Out of the question. The man stood up and shook our hands.

Walking to the door Melle asked: "Did the torpedo boats come here by train? Perhaps we can measure *them*?"

Incensed, the civil servant snapped: "Where are your passports? You said you were residents. What does this 'Haplo...' mean?"

Now, to stay calm. Take the time to explain to him in even greater detail how we had come to be in Tanzania and why the little wooden research boat didn't meet our needs. Improvise a long speech with a preamble and an epilogue. Tanzanians are mad about speeches. Ask the milkman about milk and he will talk about it for an hour and a half without a moment's hesitation: about its taste, about how to dilute it without losing the creamy quality, about stale milk that turns sour, about where it comes from and where it's going.

We finally succeeded in convincing "olive-head" that we were not trying to filch secrets out of him and that we had no affiliations with South Africa, even though we did speak "Ki-Holanzi." But it was out of the question that we measure the torpedo boats.

"So you're researchers. Good, our country needs people like you, *karibuni* Tanzania, welcome to Tanzania," said the civil servant, as if reciting a government directive.

I thought I'd try once more. "Do you perhaps know which tunnel is the narrowest, even if you don't have its measurements?" I asked carefully.

"In fact, officers, this is simple. The last tunnel is also the narrowest one. It is the very same tunnel." The man was constantly shifting between English and Swahili. "The remains of Western products that couldn't pass through the tunnel lie near its entrance." Smiling, he concluded: "So that, my dear scientists, we may call this: the law of the last tunnel."

The Narrowest Point

The following Sunday morning Melle and I walked along the railway, heading toward the narrowest point. We had no idea how far it was. An hour, a day, a week's walking? It was not marked on the map. We had a thermos filled with coffee, biscuits, mangos, rope, a folding ruler, and a knife.

It had rained. The fields on both sides of the railway were a succulent green. A soothing warm wind was blowing. Rice had been planted in the marshy areas, and maize, millet, and cassava in the drier parts. Gleaming white cattle egrets picked their way through the rice fields, hunting cautiously. A kingfisher was pecking a hollow for himself in the clay wall at the edge of a field. A bit further along, a glint of red swayed back and forth on a stalk of maize, an eye-catching red. Then it was gone. Melle balanced on the rails like a tightrope walker. I walked on the tracks behind him.

A few meters ahead of us, an old woman and a little girl emerged from a field of maize. They began walking hand in hand in the same direction as us. The woman was carrying a book on her head. She greeted us without turning around or slowing down. Did we perhaps feel like going to church with her and singing? Didn't we like singing then? It was quiet for a moment. Then she changed the subject.

Questions for the taking: "Are your parents still alive, *baba*, father? Are they healthy? Glad to hear it. Are you married?"

"*Bado*, not yet," I replied.

"Brothers and sisters? Three sisters? Very good, father," said the woman. "*Tumbo moja*, from one stomach? Not bad, four times from one stomach. Which country do you come from? A big country, I imagine. No, a small one? But with big fields, I think?"

All this time Melle and I were strolling along a few meters behind the churchgoers. The little girl had turned around a couple of times but the old woman hadn't done so once. We approached a group of baboons sitting on the tracks. The woman crouched down without removing the bible from her head, picked up several stones and threw them, with unexpected force and accuracy, at the baboons.

"They eat our maize," she said, by way of explanation. "Say, father, we're turning off here. Please say hello to your parents for me."

Without turning around, the woman and the little girl disappeared into the towering maize stalks.

Melle and I increased our pace. After walking for several hours we saw the tunnel in the distance. Two giant boulders were suspended above the rails. Formidable masses of rock on either side of the track prevented them from falling. Might this be the narrowest point? We walked under the boulders and looked around. Where were the remains of the Western donations—the carcasses of ships, threshers, cranes, and lorries? In my imagination, the mountain of rust was as high as a pyramid, but there wasn't a trace of it to be seen here. *Was* this the narrowest point? Was this even a tunnel? Melle suggested continuing to the next tunnel, saying: "Then we'll know more." I didn't think I wanted to know more. Meanwhile we walked back and forth under the boulders trying to establish which dimensions of the passage might present a problem. I looked for a sisal plant, cut off the stalk, and stretched up to reach the roof of the tunnel. Using the folding ruler, Melle measured my length, plus that of my outstretched arm and the sisal stalk: it was approximately five meters. The boat could easily pass beneath that. Then, using the rope, we measured the width: exactly 390 centimeters. I wrote down the figures in my notebook. We sat on the rails, taking turns drinking coffee from the only cup and eating a stringy wild mango.

"Let's go back. We can telegraph the facts just as they are: 'These may or may not be the measurements of the narrowest point.' We could also ride on the train and ask the driver to stop in the narrowest tunnel."

"Africa," said Melle.

I stood up and screwed the top on the thermos.

We set off for home. After a few hundred meters, a voice called up from the bottom of the slope bordering the rails.

"Wanderers, *karibuni nyumbani*, welcome to my house."

Melle walked toward the voice.

"They don't really expect us to come," I said.

It was a ritualized form of invitation. In fact, it was considered impolite not to invite passers-by into one's home. But Melle insisted on accepting the invitation. If anybody knew the narrowest point it would be these people. We crossed a narrow, sandy path and headed toward a round mud hut. Of course it was round: otherwise, enraged ancestors could hide in a corner. An assemblage had been attached to the top of the pointed straw roof: a cast-iron hook towered above bright white snail shells that had been stuck onto pointed sticks. A painted cooking vessel, turned upside down, formed the base of the sculpture.

The person who had called out, a toothless gray-haired old man with a face full of grooves like parched mud, welcomed us onto his property. A much younger woman offered us low stools, then, after greeting us shyly, disappeared into the hut. We chatted a bit with the old man. During a lull in the conversation, I introduced the subject of the tunnel: "Does freight sometimes get left behind that can't pass through the tunnel?"

The man looked at me as if I were mad and asked me to repeat the question.

"Does freight that can't pass through the tunnel sometimes get left behind?"

"Sometimes, yes ... why not? ..." he replied, hesitatingly.

"Where are the remains of the abandoned freight then?" I asked.

The old man was taken aback. He bent over to pick up a small branch and threw it into the grass beyond his plot. "The remains, the remains ... Who can say?"

"Or can it always go through the tunnel?" I continued.

"Oh yes ... of course," resounded our host. He turned his stool around and pointed toward the tunnel. "The train approaches from the distance, driving slowly, then more slowly, until it stops in front of the tunnel. Ssssss. Pfffff. Quiet. The train is quiet." The old man had become a locomotive of his own age, his arms the wheels. His feet stamped rhythmically on the ground. Tatum, tatum. "Then the train starts up again, carefully. *Pole*, *pole*, slowly, slowly. That's how it goes. See, everything is through the tunnel."

"So that's why there are no rusty ..."

"*Eh*, exactly. I'm so glad you're here. Will you stay for dinner?"

"*Sikiliza mzee*, but wait a minute, honorable old man. If I understand you correctly, everything can and cannot pass through the tunnel?"

"*Eh*," said the old man, grinning broadly. He took my arm, saying: "Everything can and cannot pass through the tunnel. You couldn't have put it better. *Wewe mwenjeji*, you are one of us. Welcome."

We excused ourselves, saying we wanted to be home before dark, and began preparing to leave.

"They're leaving again," he called to his wife.

We spoke little for many kilometers. Then Melle began complaining:

"What are we doing here? Surely we didn't come here to measure tunnels? I've been in Tanzania for six months and what have I done? Nothing. Worse still, if I make a note of the portion of the catch that disappears from the official government trawlers before they reach land, I get a death threat from one of the fellows at the factory ... It's insane. Totally insane."

"You get yourself involved in areas where money is at stake," I replied. "Think of something useless to do. That's the secret. The Tanzanians see me as a harmless idiot. A madman who counts spots."

"What is all that business about egg-spots again?" asked Melle.

"The idea came to me because I compared the eggs of *Haplochromis laparogramma* females that I had caught in two different inlets. The eggs were quite different in size. You know how habitat-restricted many of these fish are. Some species are found only near one island or one sandy stretch. Imagine that one species inhabits two different inlets and that their respective members seldom swim to the other inlet. Let's also assume that different conditions prevail in the two inlets: in one inlet, natural selection favors the production of a relatively large number of small-sized eggs; in the other, few but large eggs are favored. When females clearly prefer males with egg-spots identical to their eggs, the males in one inlet will eventually develop small egg-spots, while those in the other will develop large ones. The different characteristics of the egg-spots of the males in the two inlets could represent the beginning of a reproductive barrier, of reproductive isolation. The development of different characteristics in the eggs will be closely followed by the development of corresponding differences in the characteristics of the egg-spots. Divergence in the characteristics of the egg-spots may be the beginning of the development of two new species. The process may repeat itself endlessly, explaining the explosive speciation of mouthbrooders."[55]

Melle kept watching the birds. I wasn't sure if he heard me. "A by-product, you see, the emergence of new species as an unintentional by-product of local differences in the characteristics of the eggs. Once a reproductive barrier appears, individuals in the two inlets will become progressively adapted to their respective surroundings. Other characteristics of the populations—in addition to those of eggs and egg-spots—can, in principle, now also begin to diverge. Perhaps they will develop different mouths, different body shapes, or different retinas, adapted to the specific demands posed by life in their particular habitat. Perhaps the species in one inlet will start to live 'more quickly,' reaching the reproductive age earlier and dying earlier than before, while those in the other will begin to live 'more slowly,' becoming sexually mature at a later age, reaching an older age, and bearing a limited number of young, which are then given a great deal of attention. Emerging genetic differentiation between species in the different inlets is not necessarily obliterated by migrants who dump genes from the gene pool—packaged in either eggs or sperm—from one inlet into the other."

"That probably wasn't the case anyway, because the females had different-sized eggs. You just said so yourself," Melle responded.

"The diversity in form of all those hundreds of species—facilitated by an illusion. A world based on lies."

"Familiar territory at last," said Melle, smiling. "First, the facts, *mbwana Tesi*. Start by measuring the spots. Have you ever measured the precision of the mimesis? Does it really matter if the egg-spots resemble the eggs exactly? Won't an overall resemblance suffice? The female only needs stimulation briefly during courtship. Do females learn to react to egg-spots or is it an innate releasing mechanism? If it is, then perhaps evolution of exact mimesis is not even possible." Melle balanced, arms outstretched, on the rails. "You don't even know whether females can distinguish spots of different shapes or sizes. In fact, you don't know anything at all. First you'd better find out whether there's a direct relationship between egg size and spot size."

Melle slid off the rails. "Did you see that?" he asked.

"What?"

"That man's cap. There, wait a moment."

Bending his knees slightly, Melle whistled shrilly through his fingers and called out, shibbolethlike: "Hey you! Yes, you, with the cadaverous face. Say 'cross-eyed crustacean.'"

It was too late. The cap had disappeared into a field of maize.

"That was an SS cap. The man obviously thought it was a great cap. But how did it ever get here?"

"Fought in North Africa. One sharp gust of a desert wind and it was relieved of its SSer. Saved in the nick of time by a Berber child at play. Only the visor was left sticking out of the sand. Quickly changed heads several times after the war, moved toward East Africa. At a clothes market in Cairo ... *Heitere Tage auf braune Köpfe.** Memoirs of a cap. That's all that's missing now," I said.

We walked along the tracks silently for several kilometers before Melle started again: "Now that I've had time to think about it, what was that about ostracods, catfish, spiny eels, and snails? They too are divided into many species in several of the East African lakes. You're not going to tell me they have egg-spots as well? You should be searching for common characteristics in these different organisms. Perhaps then you could explain why new species of furu are emerging all the time."[56]

We continued walking in silence. I sensed that Melle had just said something important. It could mean the deathblow to my theory.

"And another thing: which notion of species are you using? When I hear you talking about eggs and egg-spots, it seems to me you've been influenced by Paterson's ideas. But at the same time, you don't seem to be quite free of the ideas of the maestros—Mayr and Dobzhansky—either. I think you're going to have to make a choice. Heads or tails."

What is more important to an organism: the ability to recognize whether an individual of the opposite sex belongs to the same species and therefore is a potential mate, or the ability to determine whether an individual of the opposite sex belongs to a different species and therefore is not an eligible mate? It sounds crazy, formulated in this way. One implies the other. An organism that only mates with members of its own species is at the same time avoiding crossbreeding with members of another species. Yet it is not just semantics that distinguishes the recognition and isolation concepts—as these concepts of biological species are known—from each other. Underlying the two concepts are varying opinions concerning the origin of species and the possible adaptive value of that process. But perhaps Melle was right. I adhered to the recognition concept without having totally rejected the isolation concept.[57] It was time to pinpoint the difference.

*German for "Happy Days on Brown Heads."

Isolation

Evolutionary biologists learn to live with the fact that Darwin provided an answer to almost every question they might pose. His solutions were formulated verbally, and not, as is the case today, couched in mathematical terms, but the essence can nearly always be found in his descriptions. Yet there was one important question in evolutionary biology to which Darwin paid little attention, even though the title *On the Origin of Species* suggests otherwise. He made no distinction between evolution (which he called "modification in time") and the origin of new species. But evolutionary changes—changes in the form, behavior, or physiology of organisms over a period of many generations—can also occur without the species that undergoes these changes actually splitting into two new species. Darwin concentrated on evolutionary changes in organisms *through time* and did not clearly identify the origin of species as a problem in itself.

He was aware of the striking coloration of cichlids, but did not live to witness the discovery of the large species flocks in the East African lakes. I think he would have gone there immediately given the opportunity. There is no other group of vertebrate organisms whose number of species has increased so rapidly during a specific period of time as the East African cichlids. New species are probably still emerging very quickly in Lake Victoria, as well as in other East African lakes. Faced with the improbably large number of species in these lakes, a biologist can hardly escape the question of how these many hundreds and perhaps even thousands of species emerged. Perhaps if Darwin had seen the East African species flocks, he would have paid more explicit attention to the mechanisms underlying speciation. He did show that an evolutionary line could split into two, but spent very little time analyzing how this split came about or what exactly took place during the splitting phase:

> Closely allied species, now living [in] a continuous area, must often have been formed when the area was not continuous, and when the conditions of life did not insensibly graduate away from one part to another. When two varieties are formed in two districts of a continuous area, an intermediate variety will often be formed, fitted for an intermediate zone; but from reasons assigned, the intermediate variety will usually exist in lesser numbers than the two forms which it connects; consequently the two latter, during the course of further modification,

from existing in greater numbers, will have a great advantage over the less numerous intermediate variety, and will thus generally succeed in supplanting and exterminating it.[33]

Rare passages such as these from *On the Origin of Species* suggest that Darwin was aware of the problem, but he pays no particular attention to the mechanisms of speciation. Like a born writer, he skillfully avoids the problem, or dallies with it on occasion, but nowhere does he deal with it in a way that does justice to its importance. Eighty years later, however, renowned zoologist Ernst Mayr and geneticist Theodosius Dobzhansky did.[58]

They formulated the isolation concept of biological species based on the premise that at least two species exist: each species owes its existence to the existence of at least one other species. The individuals of species A interbreed with each other but not with individuals from species B, and so forth. Why is it that individuals of different species do not crossbreed under normal circumstances? According to Mayr and Dobzhansky, such crossbreeding does not take place because of mechanisms—isolating mechanisms—that prevent genes from being exchanged between different species. Dobzhansky, who was the first to use this term, postulated that isolating mechanisms arose through natural selection for the specific purpose of preventing crossbreeding between species. Crossbreeding between different species usually means a waste of both reproductive cells and energy because hybrids are less viable and less fertile than offspring of members of the same species. Often, there will be a selection premium on avoiding hybridization with individuals of other species. According to Dobzhansky, reproductive barriers between closely related species emerged as a result of natural selection in favor of the development of reproductive barriers. During recent years, this theory has been subject to growing criticism.

There are many ways in which a reproductive barrier between different species can arise. In older species, the barrier will be solid, in other words, stratified. Mayr once used the metaphor of an obstacle course. A male and female must overcome several obstacles in order to produce fertile offspring. This may be a whole series of obstacles, particularly where old, solid species are concerned. Only when all the obstacles have been overcome will there be successful crossbreeding. If the male and female belong to different species, it is almost certain they will fail in their efforts to crossbreed long before the last hurdle has been cleared. In the case of young, closely related species such as the furu of Lake Victoria, the number of obstacles is limited. In the most extreme instance, there is only one obstacle and the reproductive barrier is minimal. Thinking along these lines, I fantasized that the youngest species of furu remained separated from each other because of the diverse appearance of egg-dummies. But it could just as easily be another signal, such as the presence or absence of a stripe on the flank or of masklike spot patterns on the head. If there is only one obstacle separating species, hybrids will occur relatively frequently.

Naturally, species do not crossbreed if their reproductive cells never meet. Hybridization does not occur if males and females of closely related species do not physically encounter each other, owing to breeding in geographically separate areas. If they do occur in the same areas, individuals of different species often remain separated because they occupy different habitats, such as close to sunlight or in shadowy areas, close to the bottom of the lake or near the surface, exposed to the wind or in sheltered areas.

Such reproductive barriers are ecological. They are not solid. If the distribution of species starts to overlap owing to a disturbance of the habitat, then reproductive barriers disappear. Hybridization will take place unless other isolating mechanisms come into effect. I was always alert to see if I could discover hybrids among the furu. It would be so logical to find them among a flock comprising such young species. But I never encountered them. This was all the stranger since crossbreeding of species of furu carried out in laboratories had produced fertile offspring.[59] Hybridization of species can also be expected among furu when their distribution begins to overlap as a result of a disturbance in the environment or when species become so scarce that individuals can no longer find mates among their own species.

A division in time, as well as in space, can also represent an ecological barrier. Many species, instead of breeding all year round, only breed during certain seasons. A difference in timing in the period during which breeding takes place can form an effective reproductive barrier between males and females of different species. Sometimes such a barrier is very precarious as only a few genes may be needed for the genetic coding of timing.

Ethological isolation—a reproductive barrier based on behavioral differences—is less vulnerable than an ecological barrier. The nature and sequence of courtship movements and the presentation of smells, sounds, color patterns, and other signals that play a role in communication between members of the same species are specific. There is only a limited chance that mating will occur with a member of another species—except where young, closely related species are concerned. Here, specific differences manifest themselves, especially in the frequency, sequence, and intensity of certain behavior. These differences are not (yet) qualitative, as they are in older species; they are quantitative.

I became convinced of this during a visit to the Gombe Stream reserve on the shores of Lake Tanganyika. In this vestigial rain forest, not far from where our forefathers first set foot aground, lived a group of chimpanzees that had been studied by Jane Goodall for more than twenty-five years.[60] Gradually, the chimpanzees had become accustomed to the presence of people. In the past, Goodall had lured them into an open space in the jungle near her camp by feeding them bananas regularly. Seated in a grassy patch near the feeding place, my traveling companion and I were talking about what had happened that day. That same morning, while watching a group of chimpanzees, we had sketched them. After the group had been at one location for a while, they decided to decamp. They left together, keeping their distance as they passed us. One male, a senior member who brought up the rear, separated himself from the rest. He walked around behind us and at one moment looked over the shoulder of my companion, then returned to the group, assuming his former position.

By this time sunset was nearly upon us. Daylight had disappeared very quickly, like a sunset in a film shown at accelerated speed. "Look beside you," said my companion in a hushed voice. I turned my head and found myself face to face with a middle-aged female chimpanzee. As soon as she saw me looking at her, she extended her right arm toward me, with her palm facing upward and her enormous fingers curled inward. Before I knew what had happened, I had reached out my right hand and enclosed her fingers in mine. Next, she drew my arm toward her and pressed her soft lips onto the back of my hand—a classic kiss on the hand. Then she released my hand. She turned around and stood there for a few moments, supporting herself on her knuckles. Eventually, she disappeared into the jungle.

This chimpanzee had greeted me using an everyday form of address. A human being is capable, without thinking, of emulating a reasonably successful chimpanzee greeting. We had the impression that my response had created no major misunderstandings. I had never consciously observed how chimpanzees greeted each other, let alone considered how I would react if I were to be greeted by a wild chimpanzee. At such a moment, time plays no role—my forefathers and hers could have greeted each other *x* million years ago when they were still members of the same species. A Chinese writer, whose name I've forgotten, once made the kind of remark that only a Chinese person could make: he was amazed by the fact that the Germans had produced such good composers. Barely out of the jungle and already producing such beautiful music. He was right. We only emerged from the jungle a short time ago, several thousands of years after the Chinese.

This encounter with the female chimpanzee demonstrates that certain elements of behavior have remained more or less unchanged between closely related species. But even in this case, the reason for this was not absolutely clear. It was also possible that there had been a convergent development in human and chimpanzee greeting behavior.

Many years ago, I remember walking up to a blackboard in an empty classroom at Leiden University where I was to give a lecture the following day. On a previous occasion, my audience had been forced to sit in the dark for an hour because I couldn't figure out how to coordinate the many light switches on the panel. I wasn't about to let it happen a second time. I flicked several switches in the hope of learning how they worked, and watched as the room slowly brightened. Three lines of text, written in white chalk, loomed up at me from the board:

DO NOT MEET
DO NOT MATE
DO NOT MATCH

It took a moment before I realized what the texts meant. "Do not meet": males and females of two closely related species do not encounter each other. Crossbreeding does not occur. "Do not mate": males and females of two species do meet but communication is so awkward that mating does not take place. And, lastly, "do not match": communication is harmonious and males and females of different species do mate. But there is literally a clash. The reproductive cells do not unite. In species with internal reproduction, the sexual organs of the males and the females fit like a key in a lock. They are made for each other.

In the case of organisms that reproduce sexually, members of the same species always recognize something in each other, even though the recognition sometimes involves no more than a chemical recognition of reproductive cells. Many aquatic organisms with external fertilization release reproductive cells into the water, after which the union of eggs and sperm is brought about by species-specific substances. Despite this, crossbreeding between males and females of different species occurs much more readily in organisms with external fertilization, such as fish, than in organisms with internal fertilization.

When all the barriers intended to prevent mating between males and females of two closely related species have been overcome, this does not necessarily mean that an effective exchange of genes will take place. Some isolating mechanisms are not effective *before* contact is made between reproductive cells but *after*. An egg cell fertilized with a sperm cell from a male of another species will usually not develop into a healthy embryo. If this does occur, there is a good chance the flow of genes foreign to that species will run aground in hybrids with reduced viability or sterility.[61]

Recognition

The isolation concept has been sharply criticized. Paterson, its chief opponent, believes that reproductive barriers between closely related species are a by-product.[57] If organisms from different populations of one species start to diverge as a result of adaptation to local conditions, this may have an effect on signals essential for the recognition of members of the same species. Consider the following hypothetical example: lizards of a certain population become grayish-black as a result of the need to be camouflaged against the barren rocks on which they lead their scavenging existence. In another population, they become green, again because of the need to be camouflaged, but this time against moss-covered rocks. It is possible that if the grayish-black and green lizards were to meet, they would have difficulty in recognizing each other as members of the same species. Differences in body color as a result of the need to be camouflaged—a selection pressure that has nothing to do with preventing crossbreeding between organisms from different populations—could, simply as a secondary effect, form a reproductive barrier.

The same model can also be applied to the different species of furu. If two furu populations occur in inlets where different light conditions prevail, their body colors may change as a result of the need to be recognized by members of their own species. If individuals from different inlets meet, it is possible they will no longer recognize each other as members of the same species. Here, speciation is a by-product, as in the egg-spots model. Evidence that speciation can develop in this way has actually been found.[62]

Paterson does not dispute the existence of reproductive barriers between males and females of different species. They clearly do exist, as everyone can see. He only refutes the existence of isolating mechanisms. Biologists reserve the use of the word "mechanism" for cases in which one can speak of a "design," and he considers this unlikely with respect to the development of reproductive barriers. How can isolating mechanisms develop if the organisms involved live in geographically separated populations and seldom or never meet? Which selection pressure is responsible for the development of a reproductive barrier in this case? A selection pressure favoring the noncrossbreeding of organisms not inclined to crossbreed anyway is not a pressure. Natural selection does not have foresight. Reproductive barriers do not emerge in advance for the purpose of arming organisms against tempting encounters that may take place with members of closely related species at some time in the future. The avoidance of hybridization between males and females of closely related species is only favored by natural selection if organisms *do* encounter each other. If hybridization is dangerous, then there is a premium on avoiding it. Only in areas where the distribution of males and females of closely related species overlaps will there be a pressure to develop distinguishing signals. Differences in body color, courtship behavior, sounds, or smells will become more specific. At least this is what the neo-Darwinian theory predicts. The signals must become specific or genes will start to "leak" from one species to the other and vice versa. This predicted phenomenon is called reproductive character displacement and examples of it exist, although surprisingly few. A well-known case is that of two closely related species of toads. In areas where only one of the two species occurs, the males whistle to their heart's content to attract females. The whistle of the males of the two species that occur only in separate areas is virtually indistinguishable. But in areas where males of both species are found, the whistle becomes more specific and females recognize the whistles of their own males.

Paterson has rightfully pointed out that the term "isolating mechanism" is poorly chosen. There is absolutely no reason to expect that males and females of the same species occurring in two geographically separated populations will develop isolating mechanisms simply for the purpose of becoming reproductively isolated. Speciation is not a goal in itself. New species emerge as the result of a passive process that is not oriented toward any goal in the future.

Another misunderstanding is that groups of organisms that produce new species in strikingly large numbers should be considered more "successful" than organisms that remain united in a single gene pool. Value judgments such as "successful" and "unsuccessful" are utterly meaningless as long as the term "success" is not defined. If "success" is used to imply that in that particular group of organisms more species emerge, then the claim is tautologous. Which is more successful: the cosmopolitan species represented by more than a million individuals or one hundred closely related species limited in their distribution to a small, closed world and each represented by only ten thousand individuals? If, on the other hand, "success" is used to mean not becoming extinct, then, in retrospect, the cosmopolitan species might turn out be the most successful of these one hundred and one species.[63]

Paterson offers an alternative for the isolation concept of biological species: the recognition concept, in which species-specific mate recognition plays a central role. The system of mate recognition is defined as one in which the behavior of males and females of one species is coadapted so they can recognize each other as potential mates. All mobile organisms have such a recognition system. The system can include all possible signals or combinations of signals. If the senses wait for the signals, they will come of their own accord: colors, movements, smells, and sounds. The isolation concept and the recognition concept can be seen as two sides of the same coin. Paterson and his followers expect wonders from reversing the coin. I don't. But although the recognition concept probably does not represent a fundamental breakthrough in the concept of speciation, it has proved refreshing to examine the problem from a different viewpoint.

How, then, do individuals with mate recognition system B develop from individuals with mate recognition system A? In other words, how do new species originate?

The Ladies' Lake

I was standing at the top of Luguru Hill with a female visitor.

"The thing I've always found most fascinating is that this species flock developed while the lake was still a basin," she said.

I pointed to the "ladies' lake," separated from Mwanza Gulf by a dense papyrus swamp. "That's not true," I replied. "The lake hasn't always been a single basin. Peripheral lakes like the 'ladies' lake' there in the distance are widespread and were probably here before as well. New furu species originate in those little lakes. They're ideal places for speciation according to the classic model."

"Then what are you doing here?" responded my guest, disappointed. "Are you here just to rubber-stamp neo-Darwinian theories?"

"No. We're here to find out what really does happen. If the solution turns out to be the classic one, then we're out of luck."

How does speciation take place according to the classic model? The idea is as follows: several individuals of one species become separated from other members of the species by an insurmountable geographical barrier. For example, a segment of land becomes divided by the formation of a river. Land organisms that can neither swim nor fly thus become isolated from members of their own species on the other side of the river. The emergence of a range of mountains can have the same effect. When local conditions differ on either side of the barrier, the individuals eventually become adapted to their own conditions. If the barrier disappears again (the river dries up, the mountains erode), then contact is eventually re-established but the individuals of both populations may have diverged to such an extent that they will no longer recognize each other as members of the same species. In that case, two new species have emerged. Divergence in the mate recognition system has occurred, with reproductive isolation as a secondary effect. Speciation as a result of a geographical barrier is called *allopatric* (from the Greek *allos patria* = a different fatherland) speciation. Species in many groups of organisms have emerged in this way, including the furu.[58]

Populations in which individuals have little or no contact with each other also occur in undivided areas. Even when there may appear to be no barriers of any kind, they may still exist. A chemical barrier can keep two populations separated as effectively as a river or mountain range. In some instances, organisms are totally restricted to a certain kind of substrate. This is clearly visible in Lake Malawi, in places where the rocky coast is interrupted by sandy stretches. Cichlids inhabiting the rocky areas seem to be almost magnetically attached to this kind of substrate. Even a patch of sand a few meters wide can form a virtually unsurmountable obstacle. If individuals on either side of such a stretch of sand become adapted in an ad hoc fashion to their respective environments, two new species may eventually emerge.

Greenwood discovered new species of furu in Lake Nabugabo in Uganda. This small lake is separated from Lake Victoria by a ridge of sand. It is the same kind of peripheral lake as the "ladies' lake." It is home to five species of furu that are found nowhere else. Greenwood believed that the closest relatives of all the species in Lake Nabugabo could be found in Lake Victoria. The species of Lake Nabugabo distinguish themselves from their sister species in Lake Victoria primarily through the color of the sexually active males. This is probably an example of classic allopatric speciation. But such an observation is never entirely certain. No one knows whether the Nabugabo species will mix again with species from Lake Victoria in the future, if the small lake is swallowed up once again by the big one.

Based on radiocarbon dating of splinters of petrified wood, Lake Nabugabo has been estimated to be about four thousand years old. The emergence of five new species over such a short period is seen as an example of rapid speciation. But it can occur more quickly—much more quickly, in fact—as will become evident later from the description of the speciation of cichlids in Lake Malawi.

Approximately fourteen thousand years ago, the water level of Lake Victoria was much lower than it is today. At one time the lake may even have dried up completely. In that case, the species flock of more than three hundred species would have emerged within a period of fourteen thousand years. But even if the lake did not dry up entirely, it undoubtedly became fragmented into small lakes and pools. According to the orthodox allopatric model, these are the best possible circumstances for the origin of species. How often had I not sat on the shore of this enormous bathtub, fantasizing that all the water had been let out and wondering where these separate little pools and lakes would emerge. Fragmentation of the lake into several smaller lakes would have created the right conditions for the emergence of the first species of furu from that one ancestral species. New species could subsequently have appeared in the little peripheral lakes such as the "ladies' lake" and Lake Nabugabo. Although rises and falls in the water level are important, equally essential is the duration of the fragmentation of the lake relative to the time needed for the emergence of two new species.

The counterpart of the model for allopatric speciation is the *sympatric* model (*sympatria* = the same fatherland). The idea here is that species may also develop in the absence of physical or ecological barriers between two populations.[64] Maynard Smith, one of the foremost mathematical biologists of our time, gave an important impetus to the development of the theory of sympatric speciation. His theories made it plausible that sympatric speciation could only take place under very exceptional circumstances. How could two new species ever develop from one parent species when reproducing individuals of both groups continually occur in the same place and encounter each other, share the same bed so to speak? What was there to stop individuals of one group from mating with individuals of the other? Sympatric speciation is common among plants and insects, which means it is not a rare phenomenon. After all, the world is populated primarily by insects. But the debate continues as to whether sympatric speciation among vertebrate organisms is more than just a theoretical possibility.

Eleven and nine species of cichlids respectively are found in two small water-filled volcanic craters in Cameroon—Barombi Mbo and Bermin. It is highly likely that all the species in these two small lakes originated from the same ancestral species. The sequence of nucleotides in the mitochondrial DNA of the fish in each of these lakes shows great similarities. The lakes are so small and the ecological circumstances so uniform that these fish live sympatrically. Even if some kind of ecological division did exist between them, individual members of the different species would still regularly encounter one other. It is therefore possible that these species originated sympatrically.[65]

It is certain that new species can emerge without a lake necessarily having become fragmented into smaller lakes. Speciation according to this pattern has been described with regard to the cichlids of Lake Malawi. It has long been thought that thousands or even millions of years are needed for the emergence of two new species. This idea is now outdated. Speciation can occur rapidly. If organisms with short generation spans are involved, the process can even take place within a lifetime. This appears to be what has happened with the rock-inhabiting cichlids of Lake Malawi.

Islands: Rapid Speciation without Geographical Barriers

There are more than five hundred species of cichlids in Lake Malawi. With the exception of four of these species, they all originated in—and are limited in their distribution to—this one lake. There is one group of approximately two hundred rock-inhabiting species that have come to be known as the *Mbuna*, the local name for these fish in the Chitonga language. A striking feature of the Mbuna is that they are poor travelers. Another unusual feature is that for fish, they lay very few eggs. In some species, the female lays no more than five or six eggs per clutch, a number common for birds or mammals but not for fish. Compared with most fish eggs, they are large—approximately seven millimeters in diameter—and they are brooded in the mouth of the female.[66]

Evolutionary biologists have concentrated on studying these fish because they are easy to observe: they seldom relocate and the water they inhabit is crystal clear. Rocky stretches along the shoreline, which are frequent but fragmented, are continually separated from each other by other habitats. There are also many rocky islands. The coloration of the males differs from island to island, showing a lesser degree of variation on the same

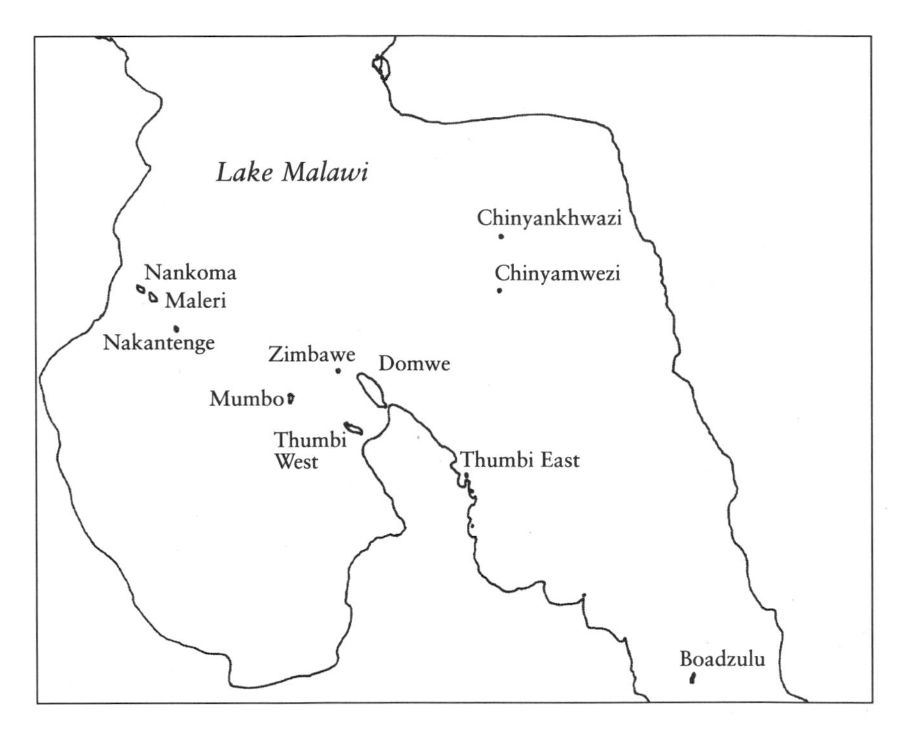

Figure 5.1
Southern part of Lake Malawi.

island. More often than not, the taxonomic status of these color forms is unknown. Do they belong to the same species, or are they (already) distinct biological species? In any case, it is certain that genetic differences between the Mbuna are minimal, as they are between the furu of Lake Victoria. It seems almost inevitable that the Mbuna have recent origins.

The more isolated a stretch of rocky shoreline or a rocky island is, the greater the number of species exclusive to it: these species are known as *endemics*.[67] A large number of species inhabiting islands far removed from the shore, such as Chinyankhwazi and Chinyamwezi, is endemic, but the greater portion (thirteen of the thirty-five) of the endemic species in the southern part of the lake occurs near the Maleri Islands, twelve kilometers away from the nearest rocky habitat. Moreover, most of the species that also occur elsewhere have different colors here. This pattern suggests that the difference in habitats has played an important role in the origin of new Mbuna species.

This situation is highly reminiscent of the distribution of Darwin's finches on the Galápagos Islands. There too, endemic species appear on different islands. However, the number of finch species (thirteen) is only a fraction of the number of Mbuna species (approximately two hundred). How can the emergence of these hundreds of species during a recent past be explained? After all, Lake Malawi is already quite old, in any event more than a million years old.

Here, as in Lake Victoria, the alternating rise and fall of the water level probably played an essential role.[67] Evolutionary biologists are particularly interested in the most recent rapid fall in the water level because of its implications for the spatial distribution of the Mbuna. Geological, hydrological, palynological, archaeological, and historical data were collected to try to establish, as accurately as possible, the most recent period with a low water level. Many of the data came from irreproachable sources, having been collected by researchers who had no notion whatsoever of the evolutionary spectacle that was taking place underwater near the rocks.

Historical data point to extremely low water levels at the beginning of the nineteenth century. In a publication that appeared in 1894, a certain Mr. Swann reported seeing giant trees in water about a meter deep along the shores of Lake Malawi. On the basis of the growth rate of trees, it was established that the water level must have been very low for a period of fifty to one hundred years in succession. This period ran from the end of the eighteenth century to the beginning of the nineteenth. Orally transmitted stories seem to confirm this. The Ngonde king Mwangonde ruled from approximately 1815 to 1835. According to one story, he walked through the northern part of what is presently Lake Malawi to a place called Mwela, to marry a woman called Mapunda.

Data from other sources confirm the hypothesis that the end of the eighteenth century and the beginning of the nineteenth were exceptionally dry. If it really is true, as Owen and his colleagues maintain, that the water level had fallen more than one hundred and twenty meters, then two hundred years ago many of the rocky islands with strictly endemic species were bone-dry. Cichlids that inhabited these exposed rocks would have suffocated, unless they had already left for wetter climes. Yet today, species that do not exist anywhere else can be found near almost every rocky island. From an orthodox point of view, the most plausible explanation for this is quite surprising: many color forms as well as biological species developed over a period of less than two hundred years. This means that speciation occurred rapidly, over no more than a few hundred generations. If rapid speciation could be expected anywhere, it would be here, in an archipelago of small islands. This expectation is based on calculations made by geneticists.

Compare two theoretical populations with the same number of members.[61] One large population comprises organisms that have continual access to each other; the other has a fragmented population with as many individuals.To increase the frequency of certain favorable yet rare mutations a much greater selection pressure is needed in the continuous population than in the fragmented population. Every tendency for a favorable yet rare gene to increase in a segment of the continuous population is hampered by the exchange of genes with individuals from other segments of the population in which the gene is rare. Mutations and rare yet favorable gene combinations have little chance of success. A large mass of individuals therefore has a conservative effect. This is why evolutionary experiments, innovations, and discoveries seldom take place among the masses.[68] As in art, the chance of true renewal is much greater in the margins, in the periphery of large populations, or in fragmented populations. The more effectively isolated organisms from different islands are, the easier it will be for the favorable yet rare genes or gene combinations to increase in frequency. Moreover, in small populations such as those of the Mbuna on the rocky islands of Lake Malawi, chance also plays a role. The smaller the population, the greater is the likelihood that the frequency with which a certain gene occurs will differ—by chance—from the average frequency achieved in a large population. This phenomenon, involving a chance shift in gene frequencies, is called genetic drift. Genetic drift—a genetic "drifting" from the norm—and natural selection have major effects on the evolution of island forms.[61]

The development of different color forms of Mbuna was not preceded by a radiation resulting in a large number of different food specialists. Nor are there any indications that the development of color forms coincided with the development of different trophic forms (see chapter 7). This suggests that food specialization is secondary, in other words, that it only comes into play after speciation has begun. This in turn means that speciation and the divergence of populations owing to adaptation to local circumstances do not necessarily go hand in hand. We found indications that the same could be true for the furu of Lake Victoria.[69] The sexually active males of different species of zooplankton-eaters have distinct color and spot patterns. Courtship takes place in geographically separate locations. They breed during periods that only partially overlap and each species has its own nurseries for raising its young. These are all strong indications that biological species are involved, young species to be sure, but species all the same, not just color forms, even though the morphological differences are minimal. Might the shapes of the mouths of these zooplankton-eaters be very different several thousands of years hence? Given the large differences between species in older flocks, it wouldn't surprise me.

What Is the Lungfish Missing?

It is striking that it is precisely the furu that develop new species so readily. The numbers of species among other vertebrates pale in the face of those of the furu. Even other groups of cichlids, such as the tilapias, are not species-rich. A living fossil such as the lungfish (*Protopterus aethiopicus)* is believed to have changed little since Lake Victoria came into existence. It has not even split into two species, even though it has spent as much time in fragmented lakes and pools as the cichlids. Spatial aspects of the process of speciation therefore cannot explain everything. It is inevitable that features of the organisms themselves also play a role. Several of these features have already been discussed: the limited tendency of many species to move about, the courtship coloration of the males, the possession of egg-dummies on the anal fin, and the small number of offspring receiving extensive parental care. These characteristics certainly do not prevent the emergence of new species. But I am concerned with something else here. The cichlids—and many others of the Labroidei suborder to which the cichlids belong—have an unusual food processing apparatus. In addition to a set of normal jaws, they have a second set of "jaws" in their throat. The profusion of cichlid species may have something to do with these pharyngeal jaws:[70] not with the fact that

Figure 5.2
Snail-crushing furu with reinforced pharyngeal jaws. Traces of snail shells are visible in the stomach. 1. Upper jaw. 2. Lower jaw. 3. Cracked snail shells.

they have pharyngeal jaws (they are common in fish) but with the specific way in which these jaws are suspended on muscles. It would go beyond the scope of this book to describe exactly how these jaws work. What is important is the unique new way in which they are suspended in the mouth and that this suspension offers new perspectives. A discovery, an evolutionary breakthrough. It could be the key innovation paving the way for the occupation of highly divergent ecological niches. Pharyngeal jaws, filled with teeth, can process food inaccessible to fish with normal jaws.

In the evolutionary sense, normal jaws and pharyngeal jaws have, inasmuch as possible, each gone their own way. They have developed independently. It is worth noting that relatively insignificant changes have facilitated the occupying of different food niches, in other words, the playing of totally different ecological roles. In each of the East African lakes, the cichlids have been responsible for nothing less than an ecological revolution, but an investigation of the background of this revolution reveals that no radical anatomical reconstruction was required for it to take place. Instead, subtle use of the ingredients already present (bones, muscles, tendons) was made. Anyone looking at these variations on a theme senses that an ingenious architect has been at work here, one who, given the spatial constraints, has exploited to the full every opportunity available to him.

The Central American cichlid *Heros minckleyi* can be found in two forms. One form has enlarged pharyngeal jaws suspended on strong muscles, with enormous flattened teeth resembling millstones. The other has normal pharyngeal jaws full of small, sharp teeth and a much less developed musculature. Intermediate forms are scarce. The diversity of forms is genetically determined. When I first peered into their mouths, I could hardly believe they were members of the same species. The differences were greater than those between certain genera of furu from Lake Victoria. Liem and Kaufman conducted experiments on food uptake and the processing of prey in both forms. As long as food was not limiting, both forms preferred soft prey to hard prey, but in times of food scarcity, the form with the enlarged pharyngeal apparatus was equipped to tap a food source inaccessible to the first, namely, snails with hard shells that had to be crushed. Why, then, do all individuals not have enlarged pharyngeal jaws if these offer better chances of survival than normal jaws? It is probably a question of a trade-off. The different parts of a living organism literally adhere to each other as a result of architectural compromises.[71] Fish will have enlarged pharyngeal jaws at the expense of something else, for example, efficiency in processing insects.

Liem and Kaufman believed that the two forms of *Heros minckleyi* might be in the first phase of a sympatric speciation process. They might eventually become two species. But how can a reproductive barrier develop between them? Young fish with enlarged and normal pharyngeal jaws emerge from the same brood. The two forms coexist randomly. It is only possible to speak of two species if fish with enlarged pharyngeal jaws only choose mates with enlarged pharyngeal jaws and those with normal pharyngeal jaws only choose mates with normal pharyngeal jaws.[72]

A related species, *Heros citrinellum,* exists that, like *Heros minckleyi,* also occurs in two forms: with and without enlarged pharyngeal jaws. When I read about this Central American species, I thought the solution to the riddle of explosive radiation had been found. In this species the possession of normal or enlarged pharyngeal jaws coincides with a difference in coloration.[54] One form is yellow, the other black. If the fish have a choice between a yellow or black mate, they choose one of their own color. This preference is not absolute, but, like every other feature, preference can evolve. This is another of the very few subjects that escaped Darwin's notice. In the course of time, the preference for a mate of the same color could become absolute, at which time forms become species.

Like a man possessed, I began searching for an equivalent of this in Lake Victoria, but my attempts were unsuccessful. Color polymorphism has been described for several African cichlid species, but in these rare cases no link has been established with diversity in the structure of the food processing apparatus. The same disappointment awaited evolutionary biologists in the other East African lakes. This polymorphic model is not applicable in general to the speciation of African cichlids.

The Night Watchman

One morning in April, as I was about to leave for the institute, I found one of the night watchmen from the mission at my door. He greeted me pleasantly, grinning. I asked why he had come. "It's that time again, father."

"Sorry?"

"You know what I mean. You're a good man. A child of God."

"To what do I owe that name?"

"*Christmassi tena*, it's Christmas again, father. I've come to collect my money," said the night watchman. "And coffee and *halfcaki*, cake. Bread is all right, too."

At that moment I remembered it was Easter Sunday. I assembled a package for him and we drank coffee together on the terrace. He told me he remembered the time before the retreat was there. There was only a bare clump of rock on top of the hill, and some trees a bit farther down. Trees everywhere. And a lot of hyenas. "I poisoned twenty-six of them."

"Twenty-six," I repeated.

"Or twenty-three. That's possible, too. In any case, there are no hyenas left now."

"Yes, there are. I heard one last night."

"A hyena? Just one, I bet. That wasn't a hyena. That was a *mchawi*, a magician. He can change himself into a hyena. There are lots of them around."

"When did you poison all those hyenas?"

"Oh, a long time ago. I'm old now."

"How old are you?" I asked.

The night watchman searched the ground for a twig. As soon as he found one, he pulled up the leg of his trousers. On his dark-brown shin, he scratched the years 1984 and 1930 one above the other, drew a line underneath them, and began subtracting furiously.

"*Hamsini na nne*, fifty-four," he said, laughing.

He stood up, threw the twig triumphantly over his shoulder, and departed.

I decided to forego sniffing formaldehyde that day, and to stay on the terrace and contemplate and relax instead. What had I learned about the furu until then and what did I still want to find out about them? My contract might not be renewed, and next year I could be sitting back home again.

There was still no detailed phylogenetic tree for the furu of Lake Victoria. But I was certain the molecular biologists would come up with one before long. In any case, it was very likely that the furu were a species flock. They all originated from a common ancestral species. It was also clear that there had been endless toying around and juggling with the elements from which furu were assembled. Anatomical innovations, such as the suspension of the pharyngeal jaws, might have paved the way for a radiation of forms. Over time, countless new forms had been generated and tested. Most of these experimental forms had been wiped out through natural selection. A small portion—still comprising hundreds of variations—continues to exist today. An exceptional example of adaptive radiation.

In contrast to what I had thought sometimes in the early phases of my research, I had also become convinced that the furu were not a single species manifesting itself in hundreds of different forms, not a super species with hundreds of masks. No, the furu were true biological species, albeit very young ones. As species, they are more closely related genetically than are Africans and Europeans, who belong to the same species.

Some of these species probably originated during periods when the water level fell and the lake was fragmented. This was speciation according to the orthodox model—classic allopatric speciation. Moreover, it seems probable that a number of species developed without the water level having fallen first. Populations near different rocky islands, in different inlets, or at different depths of the lake were sometimes extremely isolated from each other.[62] And what about sympatric speciation? I was still not convinced that this applied to the furu, although the example of radiation in the small lakes of the volcanic craters seemed to point strongly in that direction. I believed that the existence of physical or ecological barriers was a precondition for the emergence of new species of furu. The fact that biologists were unable to find these barriers did not mean they did not exist. That was about as far as I had come. But there was still so much that was unclear. Why, for example, were furu males larger than furu females and why were only the males conspicuously colored? Natural selection works along broad lines, affecting males and females of a species equally intensely. But this cannot have been the case where the emergence of these gender-related differences was concerned. If natural selection was responsible for the evolution of these differences, then it must have been a special form of selection. A form of selection that did *not* affect both sexes with the same intensity—the sexual selection that Darwin, in 1871, elevated to a category in its own right.[73] Might competition between males for females have been the driving force behind the emergence and continuation of these gender-based differences? Or was the evolution of these secondary sexual characteristics the result of choosiness among females in seeking a mate? Females that shaped males to suit their own needs? Which of these things were true? More than a century after Darwin had published his theory of sexual selection and almost half a century since Fisher had worked out Darwin's ideas mathematically, it was starting to dawn on me.[74]

Did males play a role determined largely by females? How important was sexual selection in the evolution of the furu?[54] How important was it in the emergence of the previously mentioned differences between the sexes? In the forms assumed by mating systems? In the origin of new species? Divergence of mate recognition systems in different populations of a species can occur extremely rapidly when strong sexual selection takes place.[75] There was still much work to be done. A whole group of biologists could work on these questions for decades to come, and community ecology hasn't even been mentioned. It was still incomprehensible to me how such a large number of species of the same trophic type could continue to coexist without competition for space or food having become fatal to any one of them. Was there structure in this complex furu community or did the furu coexist at random? Where to start? It was such a complicated ecosystem. A lake like a sea, as turbid as pea soup. A pea soup into which every week Mhoja, Elimo, and I dropped a net at a certain spot. A net that we dragged along behind us for a while, then hoisted aboard, and which yielded us a bucket of fish. Then I took a thermometer and thrust it deep into the lake. If I did this often enough, I would come to understand how evolution worked.

6

The Dowry: Sexual Selection and Gender-related Differences

Adults who are not continually engaged in producing offspring or totally absorbed in raising children or grandchildren have little social status in Tanzania. A childless woman in that country has virtually no status at all. When a woman bears her first child, she loses her name. From then on she carries the name of her child, preceded by "mother." If a man remains childless, he runs the risk of becoming an object of ridicule: "All that meat but no potatoes; throw it to the alligators." It is no different among the wanderers. They too, encouraged by the Africans, bear one child after the other and are welcomed joyously into the breeding masses.

Mahella, a girl who had been visiting me almost every day for several weeks, thought there must be something very wrong with me. She confided her thoughts about me to Levocatus: "That boy is either sick or he's a priest."

"Mooie bloemen, mooie bloemen."* These were the words she greeted me with when I came home, reeking of formaldehyde and suffering from a headache, after a day of classifying fish, microscopic crustaceans, and insects. She always wore the same tight dress that, by prudish local standards, had a plunging neckline. It revealed two ominous breasts. For hours she would hover around me with her full, contoured lips. Enticing. Sensual. Those lovely pear-shaped buttocks. And me suppressing my desire: look up how to say "pear" in Swahili. I almost couldn't get her to go home at night. I started admiring her: she knew what she wanted; I didn't. When I asked her if it wasn't time to go home, she began talking about danger on the road. She heard lions, I heard only nightjars. When I finally took her home, she grumbled about wanting to drive on to the Netherlands.

*Dutch for "beautiful flowers."

No, I wasn't interested. That is, I wanted sex but not a relationship. If only this woman just wanted my genes ... but I was sure that wasn't all she was after. Mahella wanted and was absolutely determined to have a wanderer. Gentle or aggressive, intelligent or stupid, fat or skinny, it didn't matter. It wasn't who she had in front of her that was paramount, but what he had to offer both her and her offspring: status, wealth, security. For months, I had been trying to establish whether female furu chose their mates with care or whether they just took the first male they came across, but as soon as the fish in the aquarium heard my footsteps, their sexual drive disappeared altogether. They waited quietly behind a stone and stared at me, biding their time until they were rid of me once more. Yet I hardly stepped through the door and there it was again: "Mooie bloemen. Mooie bloemen."

If Mahella was successful in finding a wanderer, she would have a good chance of doing well for herself, at least in the material sense. Anyone who was any kind of wanderer was a guarantee for a flood of amenities: refrigerators, freezers, cassette recorders, transistor radios, fans. A wanderer was an import office, an insurance policy, a safety net for one's offspring. With a wanderer, you flew to such exciting places as Hillegom, a small town in the Netherlands where Mahella's sister had settled with a Dutch doctor in an ordinary terraced house. Amid Holland's expansive tulip beds in an area overflowing with wanderers. It was ideal. She had been there for three weeks herself and come back with a photo album: Mahella standing next to a well-stocked refrigerator, beaming as she held the door open; Mahella bending over a stereo system, playing a record; Mahella seated in front of an electric sewing machine.

But if a Tanzanian woman had chosen a partner, and the man of her choice had been approved by her family and paid the dowry, it was better to stay away from her. "*Hodi, hodi*, anybody home?" somebody called at my door one day. I went outside and encountered a sturdy man of about forty. I had never seen him before but he said he knew me: the driver of the Land Rover who regularly gave people a lift on the road between Mwanza and Nyegezi.

I immediately fell into my role of benevolent do-gooder: "It would be silly to let people stand there for hours waiting for a bus while I drive by with an empty ... Of course I give them a lift."

"Oh yes, of course?" replied the man fiercely. "Do you have any idea what I paid for that woman?"

I had no idea who he was talking about.

"She cost me one hundred thousand shillings, wanderer. And I didn't pay it for the pleasure of seeing you drive up here with her every afternoon. Don't let me see it again!" Cursing, he stalked away, before I had the opportunity of convincing him of my innocence.

By this time, I realized who he was referring to. A woman who worked in Nyegezi and drove back into town with me every day after she'd finished work. Was I supposed to pass by her exorbitantly priced, expansive, quavering backside in the future without picking her up? It would be too painful. She was so friendly. I decided to ask her what she thought of this herself.

Time and time again, without realizing it, I had been the cause of dramas between spouses, or had spoiled budding relationships. Although the victims wouldn't be the slightest bit interested in knowing this, it taught me a great deal. I was learning as much about sexual selection from the Africans as from staring at all those fish. Only years later did I realize how tragic this next story about mate choice was.

I did my best to limit my possessions and to avoid showing them off, but with little success: a brick house, electricity, a toilet, a vehicle. These were unobtainable ideals for many Tanzanians. In my view the house was virtually empty, but the boys who helped me with my fishing gave the impression of entering a world of opulence when they came to eat with me or drink coffee. In the main room was a simple wooden table with a lamp fastened to it. On the table were a cassette recorder and a bright red biscuit tin. As it turned out, Mhoja had been secretly coveting the tin, and asked if he might have it when I left. It was only when he told me this that I even noticed the tin existed. Around the table were a few wooden chairs and two stools. The walls were bare and white. An open book lay on a wooden ledge. A print of a painting: yellow man with cart. Surely I couldn't make these fishermen jealous with a work of modern art?

Elimo sensed perfectly that I was ashamed and took advantage of it by getting me to do all kinds of little jobs for him. Recently he had asked me to go with him to the house of Maisha, the girl whose hand he had asked for.

Maisha lived in a remote village at least three hundred kilometers away. "Why do you want me to go with you anyway?" I inquired.

"Because I want to take a Nile perch with me that is as big as a goat, and a heavy bag of rice. No buses come there. Will we go in the Land Rover?"

We arranged a date and on the agreed Saturday morning, Elimo and one of his nephews appeared at my door.

That same day, after a journey of several hours, we drove up in front of a mud hut on the outskirts of a village. Maisha seemed nervous. She introduced us to a woman who asked us to sit down on a narrow bench in a dark passageway and wait. Elimo and his nephew whispered to each other. Were the fish and rice still in the Land Rover? Where had Maisha gone? I decided not to get involved.

From the passageway, part of the grounds was visible. Pieces of cassava were drying on rectangular mats. Fat runner ducks were foraging about, the males with profuse pink fleshy lobes around their beaks. They waddled toward two small ancestral huts, open structures made of branches bunched together at the top and stuck into the ground in a circle at the base. These airy forms offered shelter to the *masamva* or ancestors, who were every bit as present as the living. When I stood up to take a better look at the huts, Elimo insisted I sit down again. At last, a day off, and where had I ended up? In a dark passageway, heavy with an overwhelming odor. An all-embracing odor: the smell of firewood, perspiration, oil, cardamom, chicken, mango, and many other things I couldn't place. A map of Asia and an advertisement for a hair dryer were hanging on the mud wall. The buzzing sound of a swarm of bees approached, reached a brief climax, and receded again. I walked outside. Several stones had been pushed together near the ancestral huts. Next to them lay a mango and three limes.

When Elimo called my name, I returned to the passageway. I didn't understand the significance of the configuration of the stones and fruit, but sensed it would be better not to ask. Our hostess gave us mazabethi filled with water to wash our hands in and a girl brought us food: rice and green vegetables with chicken.

"Shall we leave after we've eaten?" I suggested, yawning.

"*Bado*, not yet," said Elimo decisively. He and his nephew ate in silence, without looking at me.

If only I were on the lake collecting furu. I wanted to get on with the egg-spots, which had almost become an obsession. Some biologists doubted whether sexual selection had played a role in the evolution of the furu. I always found that incomprehensible. One glance at an egg-spot was certain to dispel all doubt, because how else, other than through sexual selection, could these egg-dummies on the anal fins of the males have originated? It must have been the work of females who showed a preference for males whose egg-dummies most closely resembled their own eggs. But something else was going on here as well.

Male furu in deep, murky waters had egg-spots that were noticeably larger than eggs, while the spots of males in clear-water habitats were almost exactly the same size as—or smaller than—eggs. Rock-bound species in particular had small spots, while the fish themselves were large. Ethelwynn Trewavas, an expert on African fish, once published findings about the furu that she had discovered in the gut of a cormorant: they were mostly brightly colored males. Might cormorants represent a selection pressure that also played a role in determining the appearance of the egg-spots on rock-inhabiting furu? The water near the rocks is relatively clear, at least compared with other habitats in Lake Victoria. Are the species from the clear-water areas perhaps "wrestling" with a dilemma? Given the life-threatening cormorants, perhaps they would rather be without conspicuous spots, but, in order to attract females amid heavy competition, the spots are essential. Females prefer males with the largest spots. The fact that the egg-spots of the rock-inhabiting species are not larger but smaller than the eggs may be the result of two conflicting demands: being attractive to females and being inconspicuous to cormorants. A conflict between the demands imposed by sexual and natural selection, resulting in a compromise. Something similar happens with guppies. Ultimately, fish-eaters influence the color and spot patterns of guppies. If a guppy population is plagued by many fish-eaters in a certain area, the male guppies will have fewer and smaller colored spots than if the guppy population is preyed on by few fish-eaters. If fish-eaters disappear as a result of being caught, guppy spots will change as a result of the disappearance of this selection pressure. Over a period of generations, the spots will become more numerous and larger.[76]

How could I test these ideas? I could go to Japan for a year to join a community of traditional fishermen and learn how to fish using trained cormorants. Then I could return to Nyegezi to experiment near the rocks in a canoe, using tethered cormorants. Which prey would the cormorants choose from the supply available? I could paint large egg-spots on the fish in advance to see if they ran a greater risk of being caught by a cormorant because of their spots. At last I would be doing something significant. Or not? I almost forgot about the pelicans: they wouldn't waste any time in arriving. Silently, an entire group would appear, dirty plumage and all. On their way to the open water, they would pass by the rocks where my painted fish lived, disappear collectively underwater, and taciturnly scoop them up with their enormous beaks. So much for my experiments. And farewell to my trip to Japan.

The girl removed the dishes and our hostess asked us to accompany her. We walked to another hut. Again we were left in a dark passageway. About ten minutes later the woman asked me to go outside with her. Elimo and his nephew stayed behind.

A man with a stubbly chin was standing near the door of the hut and began to knead my right hand between his hands, saying "don't bother"—without my having given any indication of wanting to hold his feet.*

A group of about fifteen men were seated on the ground in the shade of a giant mango tree at the edge of the property. They were middle-aged or older. Several of them were dressed in caftans and wore crocheted caps. "Don't Bother" led me to within a few meters of them and introduced me: "This is one of our guests from Mwanza. We are very pleased he has chosen to honor our village with a visit ..."

I nodded in agreement and walked toward the group under the tree. I thought I recognized one of the men and greeted him: "Kahama!"

"Kahama ... No, I'm not Kahama," said the man.

"Oh, I'm sorry. You have a lovely place here. It's very peaceful."

The man who was not Kahama did not reply, but picked up his stick and stood up. He laid his hand on my back and led me to the spot—a few meters from the group—where "Don't Bother" had taken me earlier. He

*"*Shikamoo,*" the normal Swahili term used to greet an older person, means "I hold your feet."

looked at me intensely and added solemnly: "*Tunasumbili*, we are waiting. Our guest wishes to speak."

He bowed slightly in my direction and walked dignifiedly back to his place under the tree.

I was about to thank the men for the hospitable reception and wish them all a pleasant day when "Don't Bother" shuffled toward me and whispered: "Tell us about your forefathers ..."

My forefathers? What did he mean? What was I supposed to say? I hardly knew anything about them: Calmon, my great-grandfather on my father's side, had come from Germany to the Netherlands in 1860 and started a vinegar factory in Amsterdam. Of what interest was that to them? What was I supposed to do? Tell them about Calmon's nephew, the Talmudist, who had written a book? *Ueber das Notwendigste.** I had never laid eyes on it. The men were becoming restless. I had to say something quickly. Were the words for "eloquent" and "marrying" not the same in some languages? I couldn't think of anything to say. I didn't know where to start and thought of the past few months. Now I knew why Elimo had had those pictures taken and insisted they be in Technicolor: Elimo leaning against the Land Rover wearing a rented gold-colored watch. Now I understood the significance of the enormous fish and the sack of rice, and the purpose of my being here today: to impress.

With as many eyes as were focused on me, I now had to look back at my listeners: the eyes of my forefathers, my parents, and myself. How did I meet Elimo? What did I think of him? On what grounds did I dare to recommend him as a husband for Maisha? What could I say to further Elimo's cause? Why hadn't he warned me beforehand? He hadn't mentioned anything about what was to come. Did he think I knew? Hoping for the best, I began to ramble on about my forefathers who used to leave the wine open, in the meantime wondering desperately what I could say to recommend Elimo. Why were only the wrong things coming to mind, especially when I was so fond of him? How we went back and forth to the hospital, one time because he had the clap, another time to get a soft chancre treated in his mouth. *Could* I recommend him? That was a question I didn't want to ask myself. Several *wazee*, elderly men, stood up, nodded briefly in my direction, and left the grounds. I

*German for "On the Inevitable."

skipped more than a century and started rattling on about Elimo. About how friendly he was, intelligent, humorous, hard-working: "I have every faith in him that ..." It became noisier and the men left in increasing numbers.

This was the story of our trip to see Maisha. I'd been thinking about it again since reading a neo-Darwinian description by Monica Borgerhoff-Mulder of mate choice and bridewealth payments among the Kenyan Kipsigi tribe.[77]

In his inspiring book *The Selfish Gene*, Richard Dawkins defends the hypothesis that people, as well as other organisms, are no more than survival machines:

> We are ... robot vehicles blindly programmed to preserve the selfish molecules known as genes. ... What is a single selfish gene trying to do? It is trying to get more numerous in the gene pool. Basically it does this by helping to program the bodies in which it finds itself to survive and to reproduce. ... a gene might be able to assist replicas of itself which are sitting in other bodies. ... Are there any plausible ways in which genes might "recognize" their copies in other individuals?
>
> The answer is yes. It is easy to show that close relatives—kin—have a greater than average chance of sharing genes. It has long been clear that this must be why altruism by parents toward their young is so common. What R. A. Fisher, J. B. S. Haldane, and especially W. D. Hamilton realized, was that the same applies to other close relations—brothers and sisters, nephews and nieces, close cousins.[78]

These considerations led to the formulation of an important hypothesis: organisms do everything in their power to maximize their own chances of survival and reproduction, and those of their relatives.[79] They do not maximize their own fitness per se, but their inclusive fitness, that is, the chances of survival and reproduction for themselves and their relatives. This idea may explain altruistic behavior toward relatives. The expectation is that "altruism" will be more extensive the closer the degree of kinship.

Building on this hypothesis, Borgerhoff-Mulder wanted to test a number of specific evolutionary predictions on the Kipsigi: did men pay more cows, goats, sheep, and, since 1960, money for brides who would raise the reproductive success of the groom to a higher than average level? In the first place, the brides would be girls with high status, from families who would support their future families. Preferably girls who were sexually mature at a young age, healthy, and not promiscuous, and, most important, who had no children by other men. A woman with children by another man has indeed proved herself fertile, an important argument for choosing her, but if she is responsible for raising those children, they represent an extra burden for her future groom. He invests in raising children who are not his own, which—from the point of view of kin-selection—is to his disadvantage. Borgerhoff-Mulder discovered that men paid more for plump women than for thin ones. She suspected, like the men, that these women were healthier. More was also paid for girls who were sexually ripe at an early age. According to neo-Darwinian principles, this is correct, since these women do indeed bear more than the average number of children. More was also paid for women from afar, perhaps because in practice it was more useful for a man to have a wife who was completely devoted to him and to his household rather than one who also had to be available to serve her own family.[77]

Eventually I understood that I was supposed to act as a reference for Elimo, but the entire event was probably much more painful than I realized at the time.

While I do not know for certain whether negotiations over the size of a bridewealth payment follow the same protocol among the Kipsigi as among the Sukuma, I fear the worst: among the Kipsigi, the father of the groom pays for his son's first bride; Elimo had no parents. The father of the future groom opens negotiations about the size of the payment. He makes an offer. This offer is never accepted immediately by the bride's father. Might "Don't Bother" have been Maisha's father? Had the group under the mango tree been prominent members of Maisha's family—grandfathers, uncles, and so forth? Among the Kipsigi, the bride's father receives several candidates. If I recall correctly, there were several in Maisha's case. Then the bride's father asks a price higher than the highest bid and negotiations are held with the various candidates. Only now did I remember that one of Maisha's younger brothers had often come to visit us at Nyegezi. On each occasion, he had brought with him one or more pieces of earthenware. Unfortunately, not the beautifully designed traditional vessels but amorphous tankards. "Made in Europe," he had said laughingly, as he handed over the first two pots. During his last visit, he had asked: "Will you come to my parents' house next Sunday? Then you will sit and talk with each other and drink. Everything is 'nice and good.' Gifts will be exchanged ..." I will probably be given more of those monstrosities, I thought. I already have seven. I don't want to become a supplier of Western articles. Leave me alone. My work is continually being interrupted. I'm here to do something for the fauna, not to smuggle cassette recorders and cameras into the country for strangers. I can still see Maisha's brother's face: wincing, as if from pain, at my refusal to visit his parents. Only now did I realize that perhaps I had been chosen to act as a surrogate parent for Elimo. If that were true—and I had not yet had the opportunity of asking a Sukuma about it—then poor Elimo had been unlucky with his Wise White Employer. And he didn't get Maisha either in the end. One day, several months after our journey, he told me in a resigned voice that she had yielded dozens of cattle.

Back to the Source

My intention here was to write about sexual selection among the furu of Lake Victoria. Male furu make only a genetic contribution to the next generation, nothing more: no nuptial gifts, no territory, and no parental care. This makes them ideal fish for studying, to determine whether females choose high-quality genes festively packaged in temperamental, beautifully colored males. If a furu female is not satisfied with the first male she comes across but actually makes a choice, then she is choosing solely for his genes. After all, she cannot count on support, protection, or food. I would like to know whether the differences in color and size between furu males and females are the result of sexual selection. But this has scarcely been studied. Most furu species don't even have names yet. Slightly more is known about the cichlids from the other East African lakes, but in fact there is only one fish in which the influence of sexual selection has been thoroughly studied: the three-spined stickleback *(Gasterosteus aculeatus)*.

When it comes to behavior, a biologist can scarcely avoid the stickleback. During the past century every possible aspect of this fish has been studied. An important part of ethological knowledge is thus based on research on this fish. The stickleback has become a model organism for ethologists, like the herring gull and great tit. I, too, worked on sticklebacks before leaving for Tanzania, driven by the desire to do research in an undisturbed area. With reference to my own observations from that time and to more recent studies by others, I will illustrate several aspects of the sexual selection theory. If biologists were to concentrate on studying sexual selection in the furu, this would yield much more information than a study of the sticklebacks, but for the time being, mate choice by female sticklebacks is the best documented example of sexual selection available. So back to a small, cold Dutch lake under an interminable gray sky.

I came across the following passage in Darwin (*The Descent of Man and Selection in Relation to Sex*) on the three-spined stickleback:

> The male stickleback (*Gasterosteus leiurus)* has been described as "mad with delight" when the female comes out of her hiding-place and surveys the nest which he has made for her. "He darts round her in every direction." ... The males are said to be polygamists; they are extraordinarily bold and pugnacious, whilst "the females are quite pacific." Their battles are at times desperate; "for these puny combatants fasten tight on each other for several seconds, tumbling over and over again, until their strength appears completely exhausted."[73]

Reading this passage took me back to my days at a trout farm in the Netherlands. Near the entrance of the hatchery was a sign—I saw it again not long ago—announcing that visitors could go on a "fishing safari." The current pinnacle of Western civilization. The angler buys three live trout, which are scooped out of a fish-breeding pond and released a moment later into the sizable "safari" pond next to it. He is then free to spend half a day catching trout to his heart's content. Most of the visitors to the pond are of the male sex. Many are in poor health and go "on safari" to convalesce.

Every few hours a matronly figure passes by, pushing a cart filled with cans of beer and sausages. The "safari" hunters consume sizable quantities, as sustenance for the hunt. As soon as they are satiated, they leave their canvas chairs and set off, pacing fanatically along the shores, overgrown with reed grass and sedge, in search of the whereabouts of the trout. There, the lines of their rods find each other unfailingly and become hopelessly entangled. Flushed and cursing, the anglers pass the hours, until the lady-of-the-lamp returns to assuage their moods with more sausages and beer. This scene repeats itself day after day.

It was in this environment, kneeling at the edge of the pond, that I studied the three-spined sticklebacks, which had made their way into the food-rich waters of the trout hatchery via creeks and springs. After years of being shut away in a laboratory, a burning desire had finally become reality—out in the great outdoors, following in the footsteps of my hero Niko Tinbergen, building on the work he had started on in Leiden before the war and which is still being continued today.[80] Hadn't I been longing for this for years? Wasn't that an icterine warbler I heard squeaking in the trimmed hedge at the pond's edge? It was a perfect imitation of the cart's squeaking. Which would make the last squeak, the last warbler or the last cart? I hardly looked around, to avoid being distracted by my surroundings and to concentrate fully on the battle of the sexes taking place under the water's surface.

Close to shore, in shallow warm waters, lie the territories of the male sticklebacks, which they jealously guard against intruders such as other sticklebacks, stone loaches, and water scorpions. Occasionally, the specter of a trout looms up at the edge of a territory. Then the male stickleback recoils and appears to freeze. If a female with a stomach swollen with eggs appears at the edge of his territory, he darts over to her aggressively. A ripe female can withstand this aggression. She doesn't flee but hovers around the edge of his territory, her head raised in the "head-high" position. Then the male pricks her gently and repeatedly with his dorsal spines. After this interlude, he tries to wrench himself away again. The intention is that she wait for him at the edge of his territory while he visits the nest to make last-minute preparations. He then shoots back in her direction, jumping from side to side. If she is still ready to court, he will lead her to the entrance of his tunnel-shaped nest. Usually she follows him. Half under the female and lying on his side, he will try to entice her into the nest. If she complies, he will begin to make quivering movements with his mouth against the base of her tail, which is still protruding slightly from the nest. If she subsequently lays eggs—unlike the mouthbrooders, she lays a whole clutch at once—she will then leave the nest. The male shimmies through the nest tunnel, fertilizes the eggs, and chases the female from his territory, that is, if she hasn't already left of her own accord.[81] There is thus no pair-bonding. The male will quickly try to lead more females to his nest until it is filled with eggs. He then guards and cares for the fry for at least a week, until they leave his territory.

Sexually motivated males court not only ripe females but anything they take to be a ripe female. If they are excited enough, this can even be an air bubble! I have observed for many hundreds of hours how males try to lure females into their nests. Individual females are more difficult to follow than territorial males because they are continually on the move. They drift from one territory to the next, participating in courtships with the respective proprietors. They allow themselves to be led to the entrance of the nest and poke their snout into it. Then something strange happens. Instead of entering the nest, they usually flee and a short time later become entangled in a courtship with another male. After a while I began to recognize individual females and noticed that they continually returned to the nest of the same male. Oh no, there she is again, I would think, but the male courted her as eagerly as before. Males are prepared to accept any ripe female, but females do not accept any male. "So these women shop around for a long time before they finally make their choice," concluded an angler's wife, after I told her what I was studying. She was right.

During the same time span, a single male can be the parent of many more offspring than a single female. A female can only lay a limited number of eggs, while a male can collect many clutches of eggs in his nest and have several nests during one season. But many males are not that successful. There is a great discrepancy between the number of offspring of different males. They compete fiercely with each other for the limited number of ripe females. There is intrasexual selection and therefore a premium on a certain degree of aggression. Males who are not aggressive enough are unable to defend a large territory, and this is a prerequisite. A large territory is almost a guarantee for high reproductive success because courtship will be less frequently interrupted by rivals.[82] This form of selection to which members of the same sex subject one another has led to the evolution of protective weapons and structures (sharp horns, spines, thick skins) in many species of organisms. Males are also subject to intersexual selection: the strong selection pressure that females exert in being choosy. Sexual selection can lead to the development of secondary sexual characteristics that, except in courtship, are disadvantageous to the organism and in some cases are literally a burden. Moreover, sexual selection often clashes with natural selection in the strictest sense, as in the case of the previously mentioned guppies and the egg-spots of the furu:[76,83] sexual selection favors exaggeration of characteristics, such as conspicuous colors and sounds, pheromones, wattles, combs, and long tails, whereas natural selection in the strictest sense thwarts this development. Conspicuous features not only make a male impressive (to other males) and attractive (to females) but also vulnerable to predators. Cases in which conspicuous colors advertise poisonousness and thus make the organism less vulnerable represent only a small minority.

Female sticklebacks are discerning in their choice of mate, but for what purpose? What do they gain by not spawning in the first nest they come across? Is there a direct advantage for the female? Does her choice have direct consequences for her own chances of survival or those of her eggs? This is the case among many organisms in which the male brings the female a nuptial gift. Direct selection favoring choosiness can also be expected when males offer parental care. In the case of sticklebacks, the male not only provides parental care, he assumes total responsibility for caring for the fry. It is therefore important that the female choose a good father—one who doesn't eat her eggs or allow them to be eaten by intruders, and one who ventilates his brood well so the eggs will hatch. But how does the female stickleback recognize which male will be a good father and which will not? I will discuss the indicators of "good fatherhood" later.

Selection favoring the evolution of choosiness in females can also be indirect when the advantages of choosiness are hereditary. In this case, the female will not benefit directly from her choosiness, but her offspring will. They will inherit advantageous genetic characteristics from the male. In this case, the evolution of choosiness in females develops together with the evolution of secondary sexual characteristics in males (long tails, conspicuous colors). There has been much speculation about the possible reasons for this. Darwin, and in his wake, the mathematical biologist Fisher, gave the following explanation for the evolution of secondary sexual characteristics in males as a result of choosiness in females.[84] Females might happen to have a slight preference for males with certain external features, such as a blue body color or eyes on stalks, even though (and this is essential) these features may not necessarily be a guarantee of genetic quality. When females choose males with the most exaggerated secondary sexual characteristics, the genes determining choosiness in females and responsible for the appearance of secondary sexual characteristics in males become linked.[85] Males with exaggerated secondary sexual characteristics will produce sons with the same characteristics. Moreover, their daughters will prefer males with the same exaggerated characteristics. When, over the course of generations, the preference of females for males with extreme secondary characteristics becomes stronger, the process can run amok: the tails of male birds of paradise become increasingly long, the enlarged claws of male fiddler crabs continue to increase in size, and the eyes of male flies (*Cyrtodiopsis dalmanni)* find themselves on increasingly long stalks, until finally natural selection in the strictest sense puts a stop to an even more extreme development. This is because the moment arrives when the most extreme males are suited to little else than attracting females, and this can be fatal. The bird of paradise is condemned to the ground because of its tail, the fiddler crab topples over, and the eyes of the irresistible fly start to bounce on the ground when it alights.

If females in two populations of the same species do not have the same preference for the appearance of secondary sexual characteristics in their mates, the males of the two populations can eventually develop a different appearance. After initially being purple, the males in one area can turn blue, and those in another, red. As a result of this divergence in ornamentation, the populations can become reproductively isolated. In theory, the process can be a rapid one, particularly if the counterbalance of natural selection in the strictest sense is limited or even absent.[86] Choosiness in females could therefore be very important in the origin of new species, possibly also in the case of the furu. Some of the Mbuna species in Lake Malawi are sibling species. They resemble each other closely with regard to morphological and behavioral features, but the color and spot patterns of the sexually active males differ. Females consistently choose males with a certain color and spot pattern, an indication that biological species are involved and not different color forms of the same species. These sibling species could be young species that at a later stage will differ in more features than simply color and spot patterns.

It will not be easy to demonstrate experimentally the process of runaway male features and female preference for these features. The evolutionary

Figure 6.1
The male African fiddler crab (*Uca tangeri*). The excessive size of the enlarged claw is a result of sexual selection. (Photo Bruno Ens.)

biologist sees only the final product, not the evolutionary race preceding it, although it is sometimes possible to imitate it through artificial selection.[87]

Along our road I was continually hearing the plaints and calls and tremolos of forest creatures, but I practically never saw any of them, not counting of course the little wild pig I once nearly stepped on near one of my halting places. One would have thought from those gusts of squealings, callings and yellings that all the animals were there in their hundreds and thousands, quite close to you, around the corner. Yet when you neared the place their din came from, there were none about except those great blue [guinea fowls, *sic*], all dolled up in their plumage as if for a wedding, and so clumsy, coughing and hopping from branch to branch, that you'd have thought some accident had befallen them.[88]

This passage from Céline's *Voyage au bout de la nuit*, set in the African jungle, illustrates how striking the burden of secondary sexual selection can be. The best-known alternative hypothesis for the evolution of secondary sexual characteristics suggests that exaggerated male characteristics are always a handicap for the male.[89] When a male, despite a handicap—and the greater the handicap, the more applicable this will be—survives, this can be seen as an indication of superior genetic quality. Formulated verbally, the handicap hypothesis seems unlikely, because however suited for survival a living male with a handicap may be, without the handicap he would be that much more suited. But theoreticians have shown, mathematically, that under certain circumstances the handicap principle can have evolutionary significance.[90]

Hamilton and Zuk formulated a special version of the handicap hypothesis:[91] a female can interpret the possession of exaggerated male characteristics that are in good condition as a sign that the male is resistant to parasites, a resistance that is genetically programmed. That particular male is eligible for fatherhood because there is a greater chance that his health will be good enough for him to bring a brooding cycle to a successful conclusion and, moreover, that his offspring will be resistant to parasites.

Red, Redder, Reddest

Which criteria do female sticklebacks use in their choice of a mate? Outside the reproductive season, males and females have the same visual sensitivity to red, but in spring, when territorial males turn red, something changes in the females: their eyes become more sensitive to red wavelengths.[92] This change in retina sensitivity does not take place in the males.

This is an indication that females use the intensity of the redness of the males as a criterion in making their choice. But how can this be demonstrated? Bright red males may court more vigorously or perhaps be more pugnacious or possess some other quality valued by females that goes together with the intensity of red. Manfred Milinski and Theo Bakker conducted an exciting experiment to find out whether it was the redness or some other characteristic that attracted females. Females were placed in front of two males, each with a nest, in separate aquaria identical in size, design, and so forth. The females demonstrated a clear preference for the redder of the two males. When the females were offered the same males under green instead of white light, there was no longer any preference because the intensity of the redness could no longer be distinguished.

Females do not choose the reddest males on the basis of emotional, aesthetic considerations, but because bright red males are in better health and have fewer parasites. When the bright red males become slightly infected with parasites, their color fades and consequently the preference for them diminishes noticeably. Females use the intensity of the redness to judge the state of health of the male. The red coloration is therefore not just any handicap, but one that supplies the female with important information about the state of health of the male.[93]

Why do males plagued with parasites not have a bright red coloration? Why don't they masquerade as males in excellent health that are resistant to parasites? To deceive the other sex using the color red would be an obvious choice. But deception is costly: the production and maintenance of red pigment takes energy, and this is precisely what is lacking in the males weakened by parasites. The only guarantee of honesty in communication between the sexes is the high price of deception.

In laboratory tests on sticklebacks, the female had the choice of two males, but in nature the choice of a mate is not so simple. This is not just a problem for the sticklebacks. All females that meet a succession of males are faced with it.[94] A female may have a set of basic criteria in mind that a mate must meet, an internal standard. If she comes across a male that meets these rigid minimum standards, she stops looking. In principle, the female that functions according to a yes-no rule needs no more than a "read only" memory. Having a memory with more possibilities is only essential when a female wants to be able to adapt her decision-making criteria to suit the occasion. Another rule of thumb that females might adhere to is the following: the female compares a succession of males and makes a choice when she believes that further searching will not yield better results. The female could also assess all the males first before choosing the best one among them.[94] This last strategy has been shown mathematically to be the best, as long as the female does not have to pay a price (in terms of time, exposure to predators, and so forth) for looking for a suitable mate, but this is usually not the case.[95]

Choosers and Followers

A car driver approaches a border where there is a customs check. There are two checkpoints. One has a long line of vehicles in front of it, the other none. Which one will the driver opt for? Will he join the long line because he thinks something is wrong with the other checkpoint? Or will he try the checkpoint without a line? If he does, then he is a chooser, not a follower. This dilemma also plays a role in mate choice.

Organisms can make an active or passive choice when choosing a mate. When choosing actively, the female does not allow herself to be influenced by the choices of other females. When making a passive choice, she does. In theoretical models of sexual selection, it is assumed that females choose independently. This is an unfounded assumption. As a man with a sexual interest in women, I know that meeting a woman is much easier when you are already in the company of a woman, certainly an attractive one, then when you are alone. Experiments on this were carried out recently using guppies.[96] Female guppies are strongly inclined to choose males already in the company of a female. The female guppies were given a choice between males with and without mates. The females clearly preferred those with mates. If the mate moved to the single male, then the choosing female suddenly preferred him. These experiments show that the preference of females for certain males—at least in the case of guppies—is determined socially and not predetermined genetically.

An abstract variation is the test involving a child's carrier seat. Borrow a bicycle equipped with a child-carrier seat. Have two men begin a conversation in front of an outdoor café. One of them holds the bicycle with the child's seat. Count the number of times each man makes eye contact with a woman. En route to the next café, let the other man take the bicycle. Repeat the procedure. Women interpret the presence of the child's seat in various ways. Upon inquiring, I received the following explanations: the man (with the child's seat) had dared to commit himself, assumed responsibility, didn't leave the job of caring for the children entirely to his wife. In addition, he was perceived as less threatening.

If the chance that a female will choose a male already chosen by another female is greater than the chance that she will choose an unclaimed male (that chance could, of course, also be smaller), this is called "copying behavior."[97] A precondition is that other aspects of the behavior or appearance of the male have not become more attractive to females because of his success. Moreover, the female must be able to establish that one or more females have previously chosen that male. This is what happens, for example, when females remain in the territory of the male, visit males as a group, or, as in the case of the sticklebacks, lay eggs in the male's nest.

Stickleback females have a strong preference for males with eggs in their nests.[98] Is this because males with eggs in their nests become a brighter red or court more vigorously?[99] Probably not. It is an effect of the eggs themselves, possibly of the freshness of their smell. But of what use is it to a female to register that there are already eggs in the nest? Does she copy perhaps the behavior of other females? She could do that, in order ensure that she will produce attractive sons. Attractive fathers have a reasonably good chance of producing sons that in turn will be attractive to females. While this is possible, what really happens is probably more complicated.

The decisions females can make in choosing a mate can be rendered in a game-theoretical model. In conceiving this model, the assumption was made that females are not just fooling around, but that they make decisions with an adaptive significance. If this is true, the strategy of females must be determined by three factors:[94] the price of mate choice (energy, time, exposure to predators), the advantages of mate choice, and the ability of females to discern quality differences between males.

Consider mate choice as a game that is played by two types of females. One type is a chooser and invests time in assessing the differences between the qualities of the males, the other is a follower, who always mates—in the company of a chooser—with the same male as the chooser. Females approach males in twos. These can be two choosers, a chooser and follower, or two followers. A choosing female enjoys the benefits of choice, but she also pays the price, whether she searches for males in the company of a chooser or follower. The benefits to the follower, on the other hand, depend on the company she is with. If she is in the company of a chooser, she enjoys the benefits of choice without having to pay the price. (Remember, this is a model.) It is not relevant here to know how, in practice, a follower who goes searching for a mate in the company of a chooser manages to avoid paying the price of making a choice. If a follower is in the company of another follower, then she does not enjoy the benefits of choice. Perhaps in that case her "choice" is best described as a random one.

Stephen Pruett-Jones, who developed this model, identified under which circumstances these behavioral strategies achieved a state of equilibrium, in other words, were stable in the evolutionary sense. A strategy is evolutionarily stable if it is immune to infiltration, that is, cannot be improved or displaced by a mutant strategy or combination of strategies. If the price of mate choice is higher than the benefits, then copying behavior could be an evolutionarily stable strategy: a population consisting only of followers cannot be infiltrated by choosers. But if the benefits outweigh the risks, then a population comprising only choosers is not inviolable; the same applies to a population comprising only followers. Under these circumstances, neither choosing nor following is an evolutionarily stable strategy. A mixed strategy, a population of choosers and followers, can also be evolutionarily stable. If the benefits and risks that accompany making a choice have been identified, then it is possible to calculate which ratio of choosers and followers is evolutionarily stable. Moreover, there needn't necessarily always be two distinct categories of females. Opportunism can also lead to an evolutionarily stable situation as long as females change strategies at the right moment. The fewer the benefits to be gained from choosing, the larger the percentage of followers there should be in a population.[94]

Males that Sneak and Steal

The battle of the sexes can be very intense, more intense than the first ethologists—such as Tinbergen and Lorenz—imagined.[100] It was long thought that species had one specific tactic for finding a mate, but this is not true. Females follow different strategies in their quest for a partner. But so, too, do males. Particularly in cases where some males are much more successful than others, two or more tactics will often be found for improving reproductive success. Sometimes there are two genetic types of males, each following its own path for reproduction and unable to shift tactics. More often, a single male can choose between tactics, depending on the situation.

Stickleback males with a territory and nest that are unsuccessful in getting a female to lay eggs switch to an alternative tactic: sneaking.[101] If a sneaker sees his neighbor involved with a female, he loses his color. While a few moments earlier he was darting around in his territory, shimmying back and forth, he now hangs motionless in the water as if frozen. He then sinks, apparently lifeless, to the bottom and sneaks millimeter by millimeter toward the adjacent territory. Often he has fixed routes and takes cover behind plants or stones during his prowling activities. At a certain point he appears suddenly and swims toward his neighbor's nest entrance. The result is usually that the female flees. Sometimes, when there is a female in the nest, the sneaker burrows his way into it. He swims through the nest over the female, leaving his sperm behind.

Thus far, the sneaker's maneuvers are understandable. A male tries to carry out a fertilization that, rightfully speaking, "belongs" to his neighbor. But then he does something strange. He doesn't leave the eggs—several of which he may have fertilized—in his neighbor's nest, in which case the neighbor would have to take care of them. Instead, he tries to steal them.[80] If, despite heavy resistance from the neighbor, he succeeds, he takes the eggs to his own nest. Might this behavior mean that the sneaker is trying to masquerade as a male who has been successful with other females?[102] Is this his evolutionary response to the copying behavior of females?[103] It wouldn't surprise me—even though it hasn't been studied thoroughly—because sometimes sneakers steal eggs without having attempted to fertilize them.

Males competing for females, selective females who are very sparing with their eggs, copying behavior, alternative male mating tactics: if there were a species flock in the West like the one in Lake Victoria, dozens of biologists would doubtless be studying it, because the phenomena important for the evolution of sexual dimorphism among sticklebacks are also found among furu and other cichlid species, but then in an endless variety of forms.

Is There a Future for Polyandry?: Sexual Selection and the Mutability of Mating Systems

Sexual selection not only influences the evolution of secondary sexual characteristics. It has more far-reaching consequences. It often determines the social structure of a species. But its strength is not constant, and the social structure differs according to the intensity of sexual selection. Sexual selection determines whether organisms are monogamous or polygamous, and whether males monopolize females or vice versa.[104]

Every day as I walked to my work, I passed a tree from which dozens of spherical weaver-bird nests were hanging. Weaver birds are small finchlike birds. This particular species has a yellow body and pitch-black mask and makes an incredible racket. When I walked home again at night, the racket hadn't stopped. It must have been hell to live in that tree. Crook, who conducted a large-scale comparative study of African and Asian weaver birds, discovered that most jungle species are monogamous.[105] The appearance of male and female jungle inhabitants differs little or not at all. One pair defends a territory large enough to produce sufficient food to sustain them. But males of species from the savanna, such as the weaver birds in this tree, are polygamous and have colors that are much more distinct than those of the females. They also continually commit adultery and it would seem that the females are often raped. The noise must have some basis.

Savanna species brood during periods when seeds are available locally in profusion. They forage for seeds in groups, a tactic that probably yields better results in the savanna than foraging individually. The chance of stumbling on a place where food is plentiful is greater. Moreover, predators will be less able to approach unnoticed. Crook was one of the first to succeed in establishing a link between ecological conditions and social structure.[106] This approach later developed into a branch of research in its own right.

As mentioned earlier, the battle of the sexes is the result of an unequal investment in reproductive cells—scarce egg cells, plentiful sperm cells—and in the young produced. Essential to the nature of this conflict is whether the females represent a hindrance to increasing the reproductive success of the males, or vice versa. The more acute the shortage of one of the two sexes, the fiercer the competition between members of the other sex for the scarce mate.[107]

Sexual selection is weak among monogamous cichlids. The more polygamous the species, the stronger the selection. Generally speaking, sexual selection is stronger the more effectively a smaller portion of one sex monopolizes a larger portion of the sexually active members of the other sex. An example of this is the control that a male impala exercises over a herd of female impalas: the male has a harem of dozens of females and keeps other males away. In this case sexual selection is strong.[104]

Emlen and Oring made clear that the extent to which potential mates can be monopolized depends on ecological factors. If these have been identified, a prediction can be made as to the nature of the mating system. Many species of birds are monogamous because intense involvement is required from both parents if the young are to be raised successfully. The advantages that polygamy would bring the deserting parent do not outweigh the disadvantages that would be incurred by the loss of the abandoned brood. Polygamy only has a chance of success if a member of the polygamous sex is able to defend several mates or attract several mates, by successfully defending essential resources. If this cannot be achieved in terms of energy, then nothing more than a monogamous relationship is possible. That is, unless—as is the case in many mammals—one of the two sexes can pass the responsibility of parental care on to the other sex. In many mammals, the males desert the females, "speculating" on the strong tendency of the nursing sex to stay with the young.[104] Occasionally, the nursing sex is successful in establishing a bond with more than one mate, as the following example illustrates:

And the Hindu keeps her locked up at home—you never see her. He is everything. She is nothing. I did happen to see Tibetans unlike the normal Tibetan, who is much more refined. [They were, *sic*] remarkable, for the following reasons:

The Tibetan man was sturdy, broad-shouldered, large, not handsome, and coarse.

The Tibetan female was sturdy, large, not pretty, and almost stronger and coarser.

Of these Tibetan women—as they used to be—some married as many as five men (at the same time, of course), and in my opinion, kept them on the straight and narrow. If anyone should have been locked away, it was the husbands rather than the wives.

I saw one of these women: she was in charge of the money and gave the orders, a self-assured busybody amid strapping but docile males of 1.80 meters in height.

This description is taken from *Barbare en Asie* (1933) by Henri Michaux,[108] an astute observer and unrivaled moaner, who, during his journey through Asia, also observed polyandry among Tibetan women. Later, decades later, anthropologists with a neo-Darwinian approach studied polyandry in great detail.[109] True polyandry, in which one female has several males and different males share the same female, is scarce. With the exception of people, polyandry is found only among a few species of birds, including jaçanas. Many more species are monogamous or have polygynous mating systems (harem systems), and sometimes both males and females are polygamous. This is the case with the furu. Male furu are prepared to donate their sperm to any number of females. They are polygamous. But the females are also polygamous. The eggs in their mouths have often been fertilized by different males.

Substrate brooding and mouthbrooding:[110] these are the two most important reproductive strategies of the cichlids. Substrate brooders deposit their eggs on a stony or sandy surface. A pair guards the fry for a short while after the eggs have hatched, sometimes assisted by helpers.[111] In substrate brooders, the appearance of males and females differs little, other than that the males tend to be somewhat larger. These species have a pair bond and are monogamous. Mates only become attached after a period of getting to know each other, because a certain *compatibilité des humeurs* is a precondition for a relationship if it is to produce offspring.[112] In mouthbrooders, males and females differ greatly in appearance and behavior. They have no pair bond and are usually polygamous.

Every possible intermediate mating system that could occur between the two extremes of monogamous substrate brooders and polygamous mouthbrooders does occur.[113]

Harems and Arenas

I awoke suddenly to the sound of a dull thud next to my bed. A gecko had fallen from the ceiling and scurried furiously back up the moss-covered walls of the room. Hotel Mamba on the shores of Lake Tanganyika: had I fallen asleep the evening before with my clothes on? I dreamt. Images of that incident at the bus stop in Mwanza. The pig, the man on crutches. The bus was ready to depart. Full of passengers by European standards, quite empty by African ones. The driver waited until the bus was packed. So full that most of the passengers could only leave it by climbing out through the windows. It was illegal to drive with such a heavily loaded bus. If the driver spotted a policeman on the road, he raised his hand. A double signal. A greeting to the policeman and a signal to the passengers that half of them should lie low, so the bus would appear only half full. The roof of the bus was loaded down with luggage. Sacks full of maize, chairs, bicycles, suitcases, a wheel. A tethered pig was hoisted up by two men standing on the roof. I asked if it couldn't ride inside if I bought an extra ticket. They looked at me as if I were mad and let me know it was out of the question.

Next to the bus stood a man on crutches. He had only one leg and both his forearms were missing, ending in stublike elbows. He was talking with a group of shoeshine boys seated on vegetable crates. On the ground in front of them was a small radio, turned up full volume: the president was saying that the country wasn't doing well. Nobody took notice. One passenger, a wildly gesticulating youth, stuck his arm out of a window. He was holding an orange and made ready to throw it in the direction of the man on crutches. He called to him and as soon as the man looked toward him, let the orange fly. It shot through the air with a curve, hit the shoulder of the man, and rolled into a puddle. The man tottered, fell, and slid backward, following the orange into the puddle. Slithering in the mud, he thrashed about like a drowning insect, until one of the shoeblacks helped him up. The bus was shaking with laughter.

The driver nodded for me to get in. Should I go after all? The hardness of these people suddenly offended me deeply. Then images loomed up in my mind's eye of the thousands of fish that had struggled to catch their last breath and had been speechless in the face of my cruelty.

The bus journey lasted forever. The pig shrieked at every deep pothole. Would I be able to find Tetsu Sato? This Japanese biologist, who had been so wise as to start working in a clear lake, was making one thrilling discovery after another. He had found "swimming cuckoos" and was now planning to study them more closely: catfish (*Synodontis multipunctatus)* that allow their eggs to be taken up and hatched in the mouths of different species of mouthbrooding cichlids. Young catfish prey heavily on cichlids. Not only do they enjoy the protection of the closed mouth of the adoptive parent, they also live off the cichlid embryos they find there.[114] Sato was also studying polygamy among cichlids that brooded in empty snail shells. I hoped to be able to see these fish. To become acquainted with a new organism. Nothing could beat that!

I got up. It was turning light. I washed, dressed, and walked to the beach, expecting to find snail shell-dwellers. On the lakeshore I saw a hotel employee engaged in sweeping the sand, an unusual activity. I walked toward him and addressed him. A swept pathway made a large curve from the water up to where we were standing, at which point it merged with the tracks of a large animal.

"You're up early, *patron*, exceptionally early for a wanderer," said the man.

"Crocodiles?" I asked.

"I'm removing the traces," said the employee, laughing, "otherwise the guests will think there are crocodiles here."

"Not entirely erroneously," I said.

"There used to be many more. Now there's only one. She's called Zazie. She's okay."

I walked back to the hotel and sat on the terrace, reading about the fish I didn't dare to look for. I belonged in a library, not in the field. One of the species that broods in snail shells, *Lamprologus callipterus,* distinguishes itself from the rest in that the males actively collect and defend the snail shells.[115] The shells are essential for the females. They lay their eggs in them and it is in this protected environment that they care for their fry until they can fend for themselves. The males provide no direct parental care. They couldn't, even if they wanted to, because they don't fit into the shells. They deposit their sperm near the entrance, after which the female—sucking and fanning with her pectoral fins—ensures that the sperm reaches its destination. The males do, however, defend the area around the snail shells. There is a division of labor. Because the limited supply of shells can be transported and is suitable for re-use, competition between males assumes bizarre proportions. They plunder each other's nests and steal usable shells for their own collections. When females and their eggs or fry are found in stolen shells, the females are chased away and the fry eaten. Large, strong males manage to collect more snail shells than smaller males. There is a selection premium on a sturdy build, in contrast to the females, who, above all, must be compact enough to fit into the shells. The different demands that must be met by the bodies of the males and females respectively have led to the development of extreme differences in the sizes of the sexes: territorial males are, on average, fourteen times heavier than females, a sexual dimorphism that is even more extreme than in other harem-keeping species, such as the elephant seals.[116]

A shortage of one of the two sexes is not dependent on the relative numbers of males and females, but on the number of sexually active males in relation to the number of sperm-receptive females: this is known as the operational sex ratio. Where females are not simultaneously ripe but there is a supply of fertile females or ripe eggs throughout the year, females are scarce in relation to the number of males who are sexually active for a long period of time. Competition between males becomes intense. This can lead to a mating system in which the males do not defend the females or food sources but struggle to establish their position of dominance in relation to each other. Sexually active males concentrate in arenas. These are areas where brooding does not take place and that are used solely for the purpose of attracting and fertilizing females.[104] Bowerbirds (*Ptilonorhynchidae*), found in New Guinea and Australia, are also well known.[117] The males are exempted from nesting duties. The females build the nests and hatch the eggs all on their own. To attract females, the males build "display grounds" with bits of straw, and avenues leading up to them known as "bowers." Some species build display grounds around a young tree. Males decorate the display site with stones, moss, flowers, feathers, snail shells, and, since Western man has begun to leave his traces in the deepest recesses of the jungle, buttons, buckles, the tops of ballpoint pens, clothespins, and similar objects. The males of one Australian species, the Satin Bowerbird (*Ptilinorhynchus violaceus),* have a shiny dark blue color and do not collect objects at random with which to decorate their bowers. Instead, they are very discriminating, choosing only objects with the same blue color as their own plumage and eyes, as if making their surroundings an extension of themselves. Moreover, they paint their bowers using fruit pulp or a piece of bark, which they hold clamped in their beaks as a tool, "to act as a combination sponge and stopper."[118]

Males of drably colored species build complex structures, thus "compensating" for their drabness, while males of conspicuously colored species have tumbledown structures. The females make the rounds in areas where the males' structures are found. They judge the structures and decorations on the basis of quality and leave the males to make endless courtship displays before deciding with which male they will mate. The males are subject to strong sexual selection by the females. There is intense competition for the best places to build arenas, and males with such an arena mate relatively frequently. Moreover, they steal each other's decorations and, if possible, destroy each other's structures.[117] Collias and Collias suspect that the building of bowers began when males became exempted from nesting duties. Why that happened is a problem in itself. Natural selection appears to have rewarded the use of all objects found attractive by females. Shortly after being freed from its nest-building obligations, the male probably still had the capacity and motivation to "play" with nest-building materials. But, during the evolution of this behavior, it began collecting objects other than those needed for nest-building in order to attract female attention. The collecting of materials for the nest—now separated from its original function—became symbolic. One of the species to which this applies decorates its bower with flowers, every day replacing the wilted flowers with carefully chosen fresh ones.[117] Courtship arenas are also common among many of the species found in Lake Malawi. These arenas are divided by the males into smaller areas, in which the males receive the females. Sometimes, the males dig shallow depressions in which the females can lay their eggs; other times they build high sand castles. The criteria on which mate choice is based are not uniform. In one species, *Otopharynx argyrosoma,* females have been observed to prefer the males that occupy a central position in the arena. Only the older, dominant males can claim these central positions. In other species, females choose the males with the largest sand castles, even though they are often not the largest individuals.[54]

What takes place on the thick blanket of water lilies in the sheltered inlets of Mwanza Gulf is quite a different story. These idyllic locations are the nesting site of chestnut-brown jaçanas, with their creamy white throats and cheeks. In micro-territories, which together form the empire of one enormous jaçana matron, delicate males hatch out their eggs on beds of lilies. The polyandrous female patrols her territory on her unusually long toes. She is heavier, stronger, more dominant, and more aggressive than the males, which have little input of any kind. How did this highly unusual role reversal evolve?

As already mentioned, females are usually less inclined to abandon their offspring than males. This quality is fatal to the evolution of polyandry. Among many species of birds, parental care is shared almost equally between the parents, and occasionally, polyandry develops from such a system. If there is a severe loss of eggs and young because of an unfavorable climate or intensive predation, it can be advantageous for the male to hatch the eggs himself, that is, if the females are able to produce new eggs quickly when eggs or young are lost. This is the case with the jaçanas. The large female, exempted from the exhausting task of brooding, has become a specialist in egg-laying. Once females evolve into egg-laying machines and if they are able to produce more eggs than there are males to hatch them, then a shortage of males occurs. The roles are reversed. The males are now the limiting sex over which fighting must take place. Females compete for males, and sure enough: the females become stronger, more dominant, and more aggressive than the males.[104]

The psychologist and zoologist Barash writes:

> Among human beings, social and marital arrangements are determined by cultural and especially religious convention, without obvious regard to the regionally optimum ecological arrangement. Thus, the Moslem's maximum of four wives on the one hand, the Judeo-Christian policy of only one, on the other, may or may not be biologically sound, but both were probably arrived at for reasons other than their ecological and evolutionary utility.

This may well be true, but it does not obviate the fact that it would still be worthwhile to find out in which cases rapid cultural evolution takes the wind out of the sails of slow biological evolution, and in which cases it fails to do so.[119]

There is a great contrast between Moulay Ismail the Bloodthirsty, a Moroccan ruler known to have sired 888 children, and the five docile Tibetans who spend their lives under the thumb of a single woman. The harem-keeping ruler had a position of power that enabled him to monopolize a large number of women, but how does the Tibetan woman persuade five men to stay with her? In the first place, Tibetan men who share the same wife are relatives: brothers. They own a piece of land that produces just enough food to sustain them and to allow them to pay the necessary taxes to the landlord. When the brothers combine their efforts, they can just get by, but it is very important that the land not be fragmented. One way to achieve this is to allow only one marriage per generation. This is sometimes a bitter disappointment for younger brothers. By the time they are ready for a woman, the poor matriarch has become a wrinkled old crone. Younger brothers not infrequently disappear into the monastery to become monks.[109] Although the neo-Darwinian explanation for the phenomenon involving brothers being coupled with the same woman may not be adequate, it is almost certain that ecological factors play a role here.

In the aquarium in front of me, several *Haplochromis argens* males were courting the same female. Had these fish finally accepted my presence? The female repeatedly laid eggs in the nest pit of the same male, then left him to follow another male (who didn't even have a nest pit) to the upper half of the aquarium. There the male and female swam around in circles until the female laid an egg that sank slowly to the bottom. She then turned around suddenly and took it into her mouth. Several other fish dashed toward her. When it was quiet again, the scene repeated itself several times. Was it possible that this species no longer needed a nest pit? Was this a furu species that had freed itself from the substrate? This could open up new evolutionary perspectives. After the female laid another egg, I called Melle, who was working in the corridor of the institute. I told him of my discovery, but he wasn't enthusiastic.

"You'd better see it a hundred times first before you start claiming that this species can lay eggs in the water column. If I were you, I'd get back to doing what you're supposed to be doing. You keep starting on something new."

"Don't you think this is interesting then?"

"It's marvelous, but it won't get you anywhere."

"Some people find their inspiration horizontally, others vertically. What's wrong with that?" I protested.

"*Mbwana Tesi*, the era of the generalists is past. You have to specialize now. Sink your teeth into something and stay with it for thirty years. Perhaps you'll get there then."

How much time was there before I had to return to the Netherlands? What was it I was doing again? It was clear we were dealing with a species flock. I didn't have to worry about that. Establish a phylogenetic tree? I'd rather leave that to the molecular biologists. Find out whether foraging behavior and the structure of the food processing apparatus have been optimized in the course of evolution? It would be better if that were studied in a laboratory equipped with watertight aquaria and operational pumps. Sexual selection and the origin of species? No, those subjects were too vast to allow any kind of progress to be made within a few years. In that sense Melle was right. But there was one essential point that hadn't yet been touched upon. How could such a large number of species from the same trophic group occur in the Victoria basin? How could it be that individuals of different species that had exploited the same food source had not competed with each other with fatal consequences? With a bit of luck, I would find out.

7

The Niche: The Origin of Structure in Biotic Communities

Hundreds of tiny crustaceans lay on the dish under the objective of the microscope I'd been gazing through for months. One of the crustaceans was staring at me with a giant half-crushed eye. I made a tally in my log-book. I wanted to know how many of these cyclops were found in a liter of water taken from the area inhabited by zooplankton-eating furu. Then the light went out and the cyclops vanished into a black circular hole.

Outside, vervet monkeys swaggered through the dry grass, searching for *mapera*, a seed-filled fruit. A great tumult arose when one monkey snatched a piece of fruit from under the nose of another. Everywhere around me animals were competing for food and space, members of the same species but also members of different species. Sometimes the competition was indirect, when animals ate from the same food source and it was only available in limited quantities; other times it was direct, as in the case of these monkeys, who were fighting face to face over the same piece of food. But something unusual was happening here. Although I was able to observe these phenomena by chance, I could not find them where I sought them—in the furu. Most researchers saw what they wanted to see. Perhaps I didn't want to see? Oh yes, I had preconceived ideas aplenty; I just didn't see anything.

Should I go and measure fish until the power returned? Rows of plastic buckets were lined up against the walls, waiting, filled with the furu I had been collecting over the years. What a slaughter I had carried out. Thousands of zooplankton-eaters from about ten different species. Caught at different times of day, in different parts of Mwanza Gulf, during periods of very diverse food supply. It would be years before all the specimens were thoroughly analyzed. My fingers were already bent and wrinkled from formaldehyde. But enough of feeling sorry for myself. There were many worse things than a biologist slowly changing into a decalcified rubber animal while preserving his fish.

Whenever I coughed and tasted the formaldehyde, I imagined myself in one of Europe's natural history museums. In the catacombs of an early nineteenth-century building, I was rinsing the formaldehyde from the furu, before immersing them in alcohol—a better preservative, but often scarce in the tropics. When my eyes began stinging so intensely from the fumes that I could no longer work, I left the rinsing room to look at the vast collection. First, up a flight of cast-iron stairs to reach the ground floor. Looking up from the stairs, a whale skeleton could be seen hanging high in the air. It hung there, suspended between cast-iron galleries, a bleached god with small, elegant hands. Once bellowing, washed ashore on a beach, now mute and buried in this mausoleum. It was always deathly still in the building. Sometimes I turned on the tap just to break the silence. Or walked around a bit to put myself at ease with the sound of footsteps on the cast-iron floor. During one such excursion I, by chance, opened the iron doors of a storage cubicle and turned on the light. In the cubicle, the size of a living room, were African buffaloes, heads all pointing in the same direction, their right front hooves slightly raised. During their first step into space, they had solidified into this "natural pose." How long had they been standing there like this, a gentle, almost sheepish expression in their eyes, staring at the white wall as if the film were about to recommence at any moment?

The buffaloes had elephants as neighbors. Two bulls, flanking each other, gazing southward. And two full-grown cows, also side by side, staring northward, their tainted-pink chests dusty between their front legs. Under one of the females stood a baby elephant with empty eye sockets. A plastic bag hung around its neck with three glass eyes in it. In piles in front of the baby elephant lay pieces of savanna made of plastic, topped with a note reading: "Landscape on loan if not present here."

Come on, get on with it, the savanna wasn't my territory. I walked back to the stairwell and again descended into the catacombs. Should I have a quick look for the coelacanth? It had to be here somewhere. But where? I had passed all the jars earlier and had a good look at them. Was it perhaps in that large glass jar covered in black paint? The jar, which had been shoved aside, appeared large enough to house an adult coelacanth. I crouched down and carefully slid it toward me. The lid, whose edges had been greased, let go reluctantly. I lay it aside and stepped back. The jar contained a human head. I swallowed, coughed into my hand, and tasted formaldehyde. I thought I should close the jar but couldn't resist looking at the label first. The yellowy-white head belonged to an African male and had been in the collection for almost a century. An anatomist had started work once on the left cheek. The skin and several muscles were missing. With a long pair of tweezers I fished a transparent tube out of the alcohol containing a rolled-up piece of paper. It read: "Musculus (illegible) and musculus risorius in histology department. 712-912. JJH." The laughing muscle, cut into paper-thin slices. Had a reductionist lost his way here in pursuit of the essence of the laugh? Where had the rest of the body come to rest? Had the owner of the head agreed that his life should end in this way? Two large hands descended on my shoulders ... I turned around suddenly and stood up. Hey, Melle, you came after all.

"A delay between Mwanza and Nyegezi," said Melle, walking toward the window. "Look, lateralis. There, do you see it? The chameleon. Near the clump of grass."

Melle picked up pen and paper, a scalpel, and a laryngoscope, and retreated to his working quarters in the corridor.

I walked past the supply of fish in formaldehyde. "Bottom trawl night, 25-12-83" was marked on the bucket we were meant to begin working on now. "Christmas evening on the lake," I wrote in my notebook. "Fled from Santa Claus at the yacht club. Out in the boat with Mhoja and Elimo, to celebrate, on a clear starry night, that light never disappears here."

I turned on the tap to rinse the formaldehyde off my hands, but in vain. There was no water, only gurgles and sputters. I wandered out of the room and down the corridor to the institute's meeting room. The sign "Tanzanian Research Institute for Aquatic Ecology" hung above a notice board. The room was deserted. Several lizards slipped out through the barred window. A weasel-like animal with stripes resembling a zebra's—a mongoose—stood up on its haunches in the garden as soon as it caught sight of me and sounded the alarm in the form of a sharp, high-pitched cry. Immediately, six other mongooses, uttering the same shrill cry, came and sat on their haunches, to see what was happening. They fled suddenly and the deathly silence returned.

On the wall hung a black-and-white portrait of Nyerere. In the middle of the spacious room were long wooden tables. The accompanying chairs were pushed back neatly against the walls. The shelves of a bookcase were filled with volumes of biological works. They were all written in Russian, Korean, or Japanese, with the exception of two works: a German book titled *Siamesische Zwillinge beim kleinen Maulbrüter*,* and a stenciled essay in English, refuting the notion that cats are endowed with special telepathic abilities. Take a mother cat and her litter. Separate mother and litter temporarily by having one researcher carry the mother to the attic of a building, and another researcher the kittens to the basement of the same building. At an agreed time, the researcher in the basement pinches the kittens' tails. The researcher in the attic watches carefully to see if the mother reacts ... I returned the essay to the shelf. The contents of the bookcase amazed me. Had the primary goal of the organizers of the library been to ensure that it was absolutely useless to Tanzanian ecologists? Or were they not malicious but, rather, impractical?

I left the low concrete building and walked across the courtyard to the gate of the institute. It was closed, but as soon as I rattled on it, the *askari*, the soldier, who had fallen asleep in the little guard house, woke up. "*Ndugu*, comrade, you're leaving again?" he said amicably, as he swung open the gate. I strolled down the sandy path. After I'd walked a few hundred meters, I spotted a small van rushing toward me in the distance, enveloped in a cloud of golden-yellow dust. It was the Russians—the pilots and mechanics—from the bombers. They came here every week to bathe. The van parked as close as possible to the water's edge. The door slid open and the plump Russians emerged in their swimming trunks. They began jogging on the spot. As soon as the last one had climbed out of the van, they shot nervously across the open field and dived into the water. I waved to them from the sandy path. Did they realize that in this warm, motionless water it was almost impossible to avoid contracting bilharzia? Before long they would be sitting in their bombers, their bloodstreams infested with worms. If only you could talk with them. What would they think about the fact that the literature supplied to Tanzanian

*German for "Siamese Twins in Small Mouthbrooders."

biologists was all in Russian? I waved again. The men began to swim more closely together as if they felt threatened by some impending danger. A few moments later their pasty yellow bodies re-emerged from the water. They trotted clumsily to their little van, the door slammed shut, and off they went again.

I walked back to a little vegetable plot kept by the institute's bookkeeper. Barefoot, his trouser legs rolled up, he was digging in the ground with a spade. He turned around when I called him, then bent over and picked something up, holding his hands behind his back. As soon as we were facing each other he showed me what he'd been hiding. He proudly revealed two tomatoes: "My first harvest. They're yours." I accepted the gift and thanked him heartily. Only later did I realize the importance of that moment. It took several hours after seeing those gleaming tomatoes for my vague accumulation of book knowledge and ideas about the ecological niche to crystallize into a coherent whole.

The institute. Who worked there other than the bookkeeper and the guard? Some thirty Tanzanians: biologists and their assistants, technicians, fishermen, managers, secretaries, a telephonist, and a tea lady. And we were there as well. Our salaries were paid by the West and were about twenty times higher than those of the Tanzanian biologists. We had the luxury of being able to pursue our work, to continue being biologists. Not so the Tanzanians. It was impossible for them to survive on their state-paid salaries.

The first one to call it a day was the limnologist. She had lost hope of receiving a subsidy for the purchase of an expensive light meter she had counted on having. And she had no intention of dallying with a plankton net or dirtying her hands on board a trawler. That was not why she had spent five long years studying in America, deprived of family and friends. Did they think she had earned her doctorate so she could work for a pittance counting midge larvae in the Tanzanian hinterland? One morning, she smashed several calibrated flasks to smithereens in the laboratory and stalked out. She didn't show up at work for some time. Then, one day, there she was, shouldering a spade. She used it to dig up a plot of ground just outside the enclosure of the institute. Now, she spent her days growing tomatoes.

Her assistant, who bore the unusual name of "Party of the Revolution," had quickly followed suit. Owing to the shortage of jobs, he dug up a stony patch a bit farther along. The telephonist, too, now lived from and for his vegetable garden. For years he had guarded the telephone, dusting it and making sure no one toyed with it. When he had dialed a number, he would approach and touch the telephone with religious devotion. He had longed for the moment when there would be a voice at the other end of the line, but—to his great regret—it had not happened in years. The fates of the bookkeeper, the secretaries, the manager, the tea lady, and many others were no different. All of these people, who had fulfilled diverse functions in the institute, had given up their narrowly defined roles and, awaiting better times, sought refuge in the same niche: a life devoted to farming.

I returned to the laboratory, brooding about the convergence of niches during this difficult period. Two guards were dancing in the courtyard to Pergolesi's *Stabat mater*, which was resounding through the corridors. Melle's music. The guards waved to me as I walked past and said: *Safi sana*, very good, they sing like nuns, *kama masister*, eh." Occasionally, a goat's bleating drowned out the serene voice of the countertenor. The animal was tied to the leg of a table.

"Just traded him for two watches," said Melle, as he saw me approaching. "And I don't intend to feel guilty about it."

"Melle, have you seen all the new fields being planted around the institute? These people have given up their original roles. The differentiation of niches has disappeared," I said.

Melle lay down his scalpel and folded his arms. "Have you ever driven along the lakeshore toward Kenya? You can't believe your eyes. They're catching mountains of Nile perch there. Those sleepy little villages are changing faster than you can imagine. Hordes of people are flocking to them from the countryside. And they're all finding work because of the Nile perch. There you are digging around in that turbid quagmire in search of clues to the origin of niche differentiation, and it's happening before your very eyes."

"What are you talking about?" I said. "A village isn't the same as an inlet full of biological species? It's a nice metaphor but ..."

Melle didn't allow me to finish: "The Sukuma. Sure, they're basically farmers. And they keep a few cattle and catch the odd fish. But what's happening now that the Nile perch population is increasing? People with money are switching to fishing. For their daily work. Actually, they invest so much in nets and boats that they have no choice if they are to recover their investment. For the first time, there are professional Sukuma fishermen. But that's only the beginning. Nile perch are starting to be sold wholesale, and there are also small dealers buying fish per basket or per piece. A whole processing industry is emerging. A few years from now, you will no longer recognize the villages. Nile perch are being cleaned, smoked, fried, salted, and sun-dried ... That's what all these people are earning their living at. They're becoming professionals. Even processing swimbladders is becoming a specialty, and cooking the oil out of entrails. And you should see what these fish are attracting: boat builders, traders in firewood for smoking the fish, salt grinders, restaurant owners, shopkeepers specialized in fishing wares, bicycle repairmen. There's no end to it. If you're interested in the origin of niche differentiation, you should look there. The mechanism is actually in process."

Melle stood up to make tea. "The competition is murderous. Shrewd types from the city are making enormous profits. They buy up the best fish and sell them for export. What's left is for the local merchants, almost exclusively men. Women are having a terrible time. They're being forced into a marginal position. I think they're benefiting least of all from the perch, because they have no money. They trade cassava flour with the fishermen for the leftover specimens," concluded Melle.[120]

He threw the tea leaves into the thermos and said: "You're wasting your time, if you ask me."

"How can you say that? People with different occupations can't be compared with biological species. Niche differentiation in such a village doesn't come about because of natural selection. The changes take place over a period of only a few years. No shift in gene frequencies is involved. Before you know it, the Nile perch will have disappeared and they'll be sleepy farming villages again. The changes have nothing to do with genetic evolution. Perhaps competition between individuals in a village can be compared in some ways with competition between members of a single furu species, but if it also played an important role between members of *different* furu species, competition between the furu and the villagers can't be compared at all."

"Perhaps genetic evolution is steered by cultural evolution," said Melle.

"Wait a minute, one thing at a time. Imagine for a moment that our segment of Mwanza Gulf can be compared with a village.[68] That small body of water houses twelve species of detritus-eaters, seventeen species of zooplankton-eaters, twenty-six species of fish-eaters, and ten species of pedophages. It's like a village, with twelve bakers, seventeen grocers, twenty-six butchers, ten launderettes, and so on. A strange village, because how can there be enough economic space for so many people with the same occupation? How can they coexist without there being cutthroat competition for clients, competition that eventually leads to only a few representatives of each occupation being left? A community of farmers that changes in a few years into a differentiated community of fishermen—of course, that's worth studying. But I'm looking for something else. I think it's better to compare the furu with Darwin's finches. That's what I'd like to know—if they are an aquatic variation of Darwin's finches."

There are thirteen species of finches on the Galápagos Islands, each of which is adapted to eating a different type of food.[121] Any biologist could have predicted this variety in diet—even if he had never seen a Darwin's finch eat—on the basis of the extremely diverse shapes and sizes of their beaks. The beaks of the different species roughly resemble pliers: heavy-duty linesman's pliers, long chain-nose pliers, parrot-head gripping pliers, curved needle-nose pliers, and so on. The form of the finches' beak has evolved to suit its particular task. It was easy for me to say this now, but in fact a group of biologists had had to work on this hypothesis for several decades in order to prove it experimentally.[5] In principle, the size and shape of the beak determine the types of food a bird can eat. Whether a certain type of food is actually chosen depends largely on the availability of the food, the effort it takes the bird to digest it, and its caloric value.

Some finches live in trees and eat mostly small insects, spiders, and nectar. There is also a vegetarian finch that lives off fruits, leaves, and buds. Tree finches dig insects and spiders out of dead wood. Finches that scour the ground eat seeds that they have cracked open first. And, as with the furu, there are specialists that acquire their food in an unusual way, such as the finches on Darwin and Wolf islands that use their sharp beaks to peck open the blood vessels of seabirds, and then drink up the blood that seeps out.

The Galápagos Islands, like Tanzania, have wet and dry seasons. As the dry season advances, the finches' supply of fruits and seeds diminishes. By the end of the dry season, many finches have died of starvation.

If the diets of three ground finches (the *Geospiza* species) are compared at the beginning and end of the dry season, the following picture emerges. At the beginning, when there is still an abundance of food, all three species eat soft, easily manipulable fruits, seeds, and caterpillars. The diets overlap to a large extent and anyone who observes them only at that time of year might erroneously conclude that they all occupy the same ecological niche. Later, toward the end of the dry season, the overlap in diets is greatly reduced and each species concentrates on the type of food to which its beak is best suited. The relationship between the shape of the beak and the diet thus only becomes evident during periods of food scarcity. Only by observing the birds during these periods do you discover the adaptive value of the differences in the shape and size of the beak: these differences are known as bottleneck adaptations.

Each of the different species of finches occupies a different ecological niche. Competition for food probably plays an important role in the development of niche differentiation. But how did this work with the furu? Did the different species belonging to the same trophic group, such as the fish-eaters, each occupy a different niche? There was clearly niche differentiation *between* fish-eaters, snail-crushers, algae-eaters, and other trophic groups, but what about *within* the same group? Did the individuals belonging to the same trophic group mix randomly and play the same role in the ecosystem? Did they hunt or graze in the same way and eat the same food? If this were the case, then something unusual was happening here.

On the Run from the Sungu Sungu

One Sunday morning, a youth appeared at my door with a note from the director of the institute: "Could you please visit me in the afternoon? Best regards, in Jesus' name, Katonda." Late that afternoon, I strolled down the hill en route to Katonda, who occupied a small house near the institute. As soon as I set foot on his property, a cluster of screaming children swarmed around me. Katonda welcomed me and spoke nonstop of the drought, his ambition to enter politics, and the Nile perch question. Each time before changing the subject, he inquired whether I hadn't forgotten my promise to supply him with a four-burner gas range before I left for home.

"I have fourteen mouths to feed now," said Katonda laughingly, as a clutch of toddlers crawled all over him. They were not all his children. Some were nieces and nephews with whom he had been saddled because he had a reasonable income.

One of the children ran through the room carrying a chicken feather in his hand, crying: "I'm going to kill you, I'm going to kill you." Then he slashed menacingly past the feather with his other arm.

"Why did you want to see me?" I asked.

"Right. This is an important topic," said Katonda seriously, pursuing the conversation in Swahili: "Have you seen the Ugandans already? They've arrived. But we don't have a house for them. I've inquired everywhere, but all the institute's houses are rented and, as you can see, this room is getting rather crowded."

Katonda bent forward and clapped his hands to chase a chicken outside that had settled on the bottom shelf of the more or less empty bookcase. "You are living alone in the house on the hill. So I thought ... these people are coming to do research with you. If you want to educate them, it is better they stay with you night and day."

"I don't want to educate anybody," I replied.

But Katonda insisted: "You know the terrible history of these people. They've survived the terror in Uganda. They're counting on our help and I'm counting on yours."

"Briefly then," I replied.

That same evening the Ugandans appeared at my door. Were they supposed to be students? These men were at least ten years older than I was. Fathers who had left their wives and children in Uganda to obtain an academic degree. Their first names were English: Bill, Graham, and Francis. I only understood fragments of their surnames. I showed each of them to a room. They unpacked their luggage and returned to the central room, which from that moment on became a communal area. We drank coffee and then withdrew. I lay down on my bed and reflected on the day with satisfaction: at last, people in hiding. But Francis's comment puzzled me:

"Amin was not that bad." How could he say such a thing? And Bill and Graham had not refuted it. Was I so ill-informed? Or had Katonda foisted three accomplices of the enemy on me? No wonder they hadn't been able to find a place to stay. I hardly slept that night.

The next morning I was awakened at sunrise by my guests, who had discovered a cassette tape with music by Archie Shepp, and were playing it loudly. I got out of bed and found them in the communal area having breakfast. They were drinking tea and eating the bread they had brought with them. They invited me to join them. When I grumpily snorted that I didn't intend to get up until seven o'clock, Graham replied laughingly: "Strange." I turned the music off and returned to my room, laying wide awake on my bed until seven. By the time I got up, the Ugandans had left. They didn't show up at the institute that morning. Did they need my help after all? At lunchtime I walked up the hill again to see if they had returned.

When I arrived home, the front door was wide open and I found an unfamiliar face in the communal area: that of a corpulent man. We stared at each other in silence for a few moments. The man plucked at his moustache.

"What are you doing here?" I began.

"I work here. I'm the cook, *Mpishi*," said the man, in a friendly tone.

"That's new to me. This is my house."

"Welcome," said the man. "Have a seat."

"Where are the Ugandans?" I asked the cook, who picked up an apron from the table and tied it around his waist.

"The Ugandans?" he repeated. "Prisca, there's a wanderer here at the door. He wants to know where the Ugandans are."

A young woman came out of the bathroom, singing, clad in a sleeveless, bright green dress, hands covered in suds.

"*Wamekuenda mjini*, they've gone into town. Too bad. You've just missed them."

"Do you also work here?"

"Eh," the woman replied cheerfully. "*Nimepata kazi*, I have a job."

"I live here," I told her, as my eye caught sight of a calendar print of a Swiss alpine meadow. It hadn't been hanging there in the morning.

"All the better," said the woman. "If you have anything that needs washing ..."

"Where's the key?"

"The cook has it," said the woman.

I sat down again and sighed: "All right. When in Rome, do as the Romans do."

"Eh, sir," said the cook approvingly. "Welcome to our table."

He placed a plate of steaming cooked bananas and brown beans in front of me.

"Well, at least it smells good," I said, trying to figure out where I could escape to for the next while. The lake perhaps, with Mhoja and Elimo. They could be called up at any time for military duty in the Sungu Sungu, a traditional people's militia. This vigilante group had become active again lately, because for a long time the official police had had no control over cattle theft and other crimes in the villages. Sungu Sungu chiefs wearing flat headdresses covered in chicken feathers could appear at the door at any time to recruit them for an unlimited period. Mhoja and Elimo had no desire to be enlisted and had asked me if they might not move in with me for a while.

When I told Mhoja and Elimo about my plan to stay on the boat for a few weeks, they had no objections. There was only one problem. It was Ramadan. They were not allowed to eat or drink anything during the day. It was tricky in this heat. They were already starting to feel weak and the period of fasting had just begun.

"And if I do the cooking and make sure you have cigarettes and coffee in the evening?"

"*Sawa, sawa*, sure, sure, when do we leave?"

"As soon as I've drawn up a timetable for the experiment."

The next day I made preparations for the journey. I stocked up with enough food and coffee to last us for at least a week and bought a camping stove, charcoal, and matches. I was rather overwhelmed by these activities. What was it I was looking for again?

Follow a random food chain. Algae get eaten by algae-eating crustaceans of microscopic dimensions. These freely drifting crustaceans get eaten by plankton-eating insect larvae. Insect-eating furu eat these larvae. The insect-eating fish get eaten in turn by catfish. Five trophic levels and you have reached the end of the food chain. Sometimes food chains are shorter, such as the one involving organic waste, waste-eating furu, and furu-eating catfish. But they are seldom longer. More than five trophic levels in a food network is exceptional. A species-rich ecosystem always implies a large number of species in each trophic level and a limited number of trophic levels, since a large number of trophic levels is impossible energy-wise (see chapter 9).

But how many species could a stable ecosystem support in each trophic level? Would the furu community in Mwanza Gulf ever reach its limit? This had occurred to me again recently when I was in the Serengeti.

The day had drawn to a close. A vulture glided high overhead, writing letters in the clear blue sky, a musical rendering, until making a long slash through the air, diving sharply earthward, to disappear in the grass, with me in pursuit in the Land Rover. I found a white-headed vulture a short distance from the still-steaming cadaver of a gnu. Spotted hyenas were rooting in it. Their bloody heads sank deep into the dead body, reappearing briefly every now and again. One of them looked in my direction and then upward. More vultures were arriving. Large and small. Some with bald, sweaty, greenish-blue necks. Various species: numerous white-backed vultures, a few Rüppell's vultures, and an Egyptian vulture. No Nubian vultures this time. Golden-brown jackals tripped back and forth behind the hyenas, mouths watering, not yet allowed to approach the corpse. In addition to the vultures, three marabous and a raven were standing by watching. They were looking in the other direction as if wanting nothing to do with this bloody feast.

A steaming cadaver. No, I won't start from the beginning, but from the moment the hyenas and most of the jackals were satiated and departed. The body was a carcass, stripped of most of its flesh. This was what was left for the vultures and a few other carrion-eating birds. They would have to make do with it. Six species of vultures whose distribution throughout Africa overlaps considerably. What actually took place the moment the vultures started eating? Did they all descend on the carcass at the same time? That's what I saw, but that's not what really happened.[122] In fact, there was a kind of sequence, a division of labor. The white-backed and Rüppell's vultures, with their long necks, were the first to eat. They removed large pieces of soft flesh from inside the body. A bald neck was ideal for probing deep into the cadaver and, moreover, their skulls and beaks were particularly well suited to pulling. These birds were not built for twisting or tearing off strips of meat and skin. The Nubian and white-headed vultures, on the other hand, were. Finally, there were the hooded and Egyptian vultures, which played yet another ecological role in scouring the cadaver: they picked up the small pieces of meat left scattered around it. Here, too, there was a sort of division of labor. The Egyptian vulture, like a chicken, concentrated on picking pieces of flesh from the bones, while the hooded vulture collected pieces that had fallen onto the ground. Often, long after the other vultures had disappeared, the hooded vulture would still be stalking the carcass.

Six species of vultures all living off the same food source, but specialized in processing different parts of the carcass. It was crowded around the carcass and there was a lot of aggression. The fights were usually between members of the same species but sometimes they were between members of different species, particularly if they lived off the same food source. Had the limit been reached in the number of species of vultures that could live off carcasses? Or was there room for more species, if they were to appear? I could not imagine it in this case, but it might well be true of the furu. Not only was the number of species of the usual trophic types such as the fish-eaters, plankton-eaters, and waste-eaters unusually large, but these same fish created all kinds of niches that an ecologist could never have anticipated beforehand: snout-engulfing pedophages, cleaners, scale-scrapers. Whose idea was the launderette? As creative as the conceiver of that idea was *once*, the furu were *repeatedly*, naturally without being aware of it. Ecologists could seldom predict how many more species could be accommodated in an animal community and they were certainly unable to do so in the case of Lake Victoria. As yet, I had absolutely no idea why there were only three phytoplankton-eating species but more than one hundred fish-eaters. Usually there were noticeably more species at the beginning of a food chain than at the end, but in the case of the furu, it was exactly the opposite.

There appeared to be ecological laws that determined, to some extent, the structure of cichlid communities.[123] But our understanding of these laws was obscured by various other factors. A comparison of the species flocks of the different lakes showed us that these laws did exist. Roughly the same trophic groups, such as algae-eaters, zooplankton-eaters, insect-eaters, and fish-eaters, evolved time and time again. Even unusual specialists such as the pedophages had evolved in both Lake Malawi and Lake Victoria. Evolutionary biologists, therefore, reacted with amazement when Greenwood announced that there were no zooplankton-eaters in the species flock of Lake Victoria. It was an interesting prospect: a trophic group that had evolved according to the same pattern except in that one lake. It later turned out that there were indeed zooplankton-eaters, in fact more than twenty species of them. Greenwood had looked mainly in the inshore areas, while most of the zooplankton-eaters lived in the open water.

Where a species did and did not live was determined in the first place by the nature of the environment: the physical conditions, the climate, and the availability of food. Moreover, the presence of other organisms, such as potential competitors and predators, could also be of great importance.[124] Did competition for space and food occur when the number of species tapping the same sources of livelihood became too large? If this were the case, then I should be able to find evidence of competition-avoidance mechanisms, particularly if the species resembled each other so much in appearance and behavior that they played an almost identical role in the ecosystem. During the 1930s, the ecologist Gause had come to the conclusion, on the basis of laboratory experiments, that two species depending on exactly the same food sources for survival could not coexist[125]—that is, if the sources of livelihood were limited. Eventually, one of the two species would win the struggle and be the sole survivor, in accordance with the principle of competitive exclusion. What if we applied Gause's hypothesis to Mwanza Gulf, where some trophic groups were represented by dozens of species? A researcher is trained not to foster hope—except where finding the truth is concerned—but I secretly looked forward to the moment when the principle of competitive exclusion would be shattered with a counterexample. A blow to the ecological establishment, which clung so tightly to a handful of scanty ecological principles, among which is this hypothesis. Even if I did come across competition or evidence for the existence of competition-avoidance mechanisms, unfortunately I wouldn't have discovered anything new. It would presumably mean that the mechanisms that gave structure to the furu community were no different from those anywhere else. The species flock of Lake Victoria could serve as a model for the origins of differentiation in many older and more advanced radiations, such as those of the marsupials in Australia, the lemurs in Madagascar, and the armadillos in South America.

These thoughts ran through my mind as I stirred a pan of maize porridge. It was supposed to become as stiff as putty. A few greens, a small onion, several tomatoes, followed by a tug on the dragnet to catch a fish. I crushed a few coarse grains of salt with a hammer on the thwart of the boat. The charcoal burner on which I was preparing the meal was positioned on the floorboards. I was crouched beside it. The meal had to be ready exactly half an hour after sunset.

"And what do you think of that? That the Ugandans have hired a cook and washerwoman for my house without my knowing about it?

"Normal," said Elimo flatly.

"You can't do anything without a cook," chortled Mhoja.

Elimo mended a hole in the gill net we were to use later that evening. Mhoja, forever the helmsman, stared blankly ahead. We sailed slowly out of Nyegezi Bay. Little fires lit up on the shore between the shrubs. This was where they brewed *moshi*, an illegal alcohol.

"What kind of fish would you like to eat?" I asked.

"*Sato*, Nile tilapia," said Mhoja and Elimo, without hesitation.

"That's difficult. We'll have to pull up close to shore. Over there perhaps, above the sand, past the rocks."

Mhoja prepared the dragnet and cast it out.

Whatever was I doing? I'd been on the verge of discovering something, and here I was running a floating restaurant. Mhoja and Elimo looked westward. A red ribbon of light, the border between sky and lake, became smaller and smaller. Mhoja and Elimo rattled on in Sukuma and lit a cigarette. They inhaled deeply and lay back. Sighing with pleasure they exhaled large clouds of smoke.

A short time later, as we were hauling in the net, Mhoja exclaimed: "*Mawe*, stones. We've caught stones."

He shook the death trap empty and jumped aside to avoid the catch falling on his toes. "Huh, what's that?"

In the beam of light that I shone on Mhoja with my flashlight, several small silver fish floundered on a plastic bag intended for cement and tied shut with string. Mhoja removed a penknife from his trouser pocket and cut the string. In addition to being tied shut, the bag had been firmly sealed with wide tape. Mhoja made a hole in the bag, tore it open, and shook out its contents. The bottom of the boat was strewn with bank notes. Hundreds of hundred-shilling notes.

"Uh, uh, uh." We stared at the money, flabbergasted.

"Maisha," said Elimo.

"Esther," said Mhoja.

A string of girls' names followed in whispered succession, a cross between antiphonal singing and prayer: "Salome, Rebecca, Theresa." Meanwhile, Nyerere stared up at us with a thousand smiles from the depths of the boat.

A moment later, while rummaging around in the bank notes, Elimo said: "*Haina maana*, it has no value."

"Old money," replied Mhoja. "Do you think we can still redeem it?"

"No," I answered. "That was only allowed for a short while. To prevent money from being laundered. This money is worthless."

I cleaned several furu and boiled them.

Mhoja and Elimo returned the bank notes to the bag, grumbling to me all the while in Swahili: "And who had to line up in the bank to convert the money for the Indians? Exactly. We did. In small portions, so no one would notice."

"Let them launder their own money. But do you think they'd dare? No. They're too afraid of getting caught."

"*Wabaya*, bad people," concluded Mhoja, shaking his head.

"And responsible for the collapse of the Weimar Republic," I muttered, piling the maize porridge onto the Wedgwood dishes I had once bought at a street market in Amsterdam.

The Experiment

I had chosen zooplankton-eaters as the subject of my experiment. There were at least fourteen species of them in Mwanza Gulf. Half of them were easy to catch. They were ubiquitous. Moreover, they were small and fitted nicely into your hand. But the most important reason for choosing them was that they resembled each other so closely in body plan. Judging by their appearance they played a similar, if not identical, role in the ecosystem. If *these* species did not compete with each other for space or food during periods of food scarcity, which ones did? If I found no evidence for the existence of competition or competition-avoidance mechanisms, I would be inclined to think that competition had played and continued to play no role in the development and maintenance of structure in the furu flock. But I'm getting ahead of myself here ... After all, was there structure, or were all the members of these species just swimming around at random, mixing with each other indiscriminately like lottery tickets thrown into a hat?

It soon became evident that zooplankton-eaters were not distributed evenly throughout Mwanza Gulf. One species lived only above sandy bottoms, the others only frequented muddy bottoms. There was also a kind of horizontal division between the mud-loving species themselves. Some appeared primarily in sheltered inlets, others in the open waters of the gulf, which were more heavily influenced by the wind. There was also one group that inhabited only deep water. A deep lake is like a tall forest. Unlike the savanna or tundra, it has a third dimension, the vertical one, throughout which species can spread. In Mwanza Gulf, the zooplankton-eaters did just this. If a vertical gill net were to be sunk like a screen, spanning the entire water column from top to bottom, some species would always be found at the same height in the net. One such species was *Haplochromis argens*, a small silvery-white fish with a red tail, which was so faithful to the upper two meters of water that magnetic forces seemed to be involved. Other species resided almost exclusively at the bottom. There was a distinct vertical layering, reminiscent of the layers of trifle. But aside from the habitat-restricted species, there was also the migrating type. In the morning they would slowly sink down from the surface to deeper waters, migrating upward again after sunset. They pursued their prey—which followed a similar trajectory—at close range. The different species of zooplankton-eaters were separated from each other in space and each occupied its own microhabitat within Mwanza Gulf, with the exception of two species: *Haplochromis heusinkveldi* and *pyrrhocephalus*. These had a more or less similar distribution, but despite this, were ecologically totally isolated because of living off a different diet, raising their young in different places, and brooding at different times.[126] Niche differentiation therefore clearly existed within the group of zooplankton-eaters, even though, at first glance, the anatomical differences between the different species appeared minimal. In each species, food preferences, habitat, and foraging techniques were combined in a unique way. The same kind of differentiation was found in each of the other trophic groups, such as the fish-eaters, snail-crushers, insect-eaters, and waste-eaters. Every time the different species of a specific trophic group were compared in detail, each species was found to occupy its own ecological niche, however limited the available ecological space.[127] A close examination of the functional anatomy of the fish also revealed considerable differences, for example, in the structure of their gills and retinas.[128]

Why was it that a species such as *Haplochromis argens*, which inhabited a clearly defined area, so seldom left its microhabitat? It was probably because the upper few meters of water represented the optimal habitat for this particular species. They were the only fish capable of foraging as efficiently during times of relative food scarcity as during times of abundance. They were quite distinct from specimens of other zooplankton-eating species caught in the microhabitat of *Haplochromis argens*, that is, in the margins of their own habitat—an indication that *Haplochromis argens* was superior in its own niche. Furu species that were superior in their own niche ... was that surprising? Wasn't it logical? It was, for those who believed that competition was the driving force behind the origin of divisions into microhabitats. But I found no evidence of competition for food or space. I searched endlessly for shifts in the vertical distribution of zooplankton-eaters that could possibly be attributed to competition for space with zooplankton-eaters from other species. Was it not conceivable that when one particular species was present in large numbers at a certain depth, another species—a potential competitor—might be driven from that area? On numerous occasions I cooked for Mhoja and Elimo in the little boat, on numerous occasions we put out our vertical nets and spent twenty-four hours on the water, but never did these excursions produce any evidence of the existence of competition for space or food. Of course not, said the orthodox ecologists, who were convinced that competition had taken place in a distant past.[129] Anything that was really important, such as avoidance of competition, had been taken care of long ago. It would be very coincidental if a biologist just happened to witness it now.

And, what was the strict separation in space, other than a competition-avoidance mechanism? The orthodox niche theory had been severely under attack since 1977.[130] Some biologists doubted whether competition for space or food had been a dominating factor in the structuring of biotic communities. They spoke of the "ghost of competition past." Such critics of the competition theory maintained that the current patterns of distribution of organisms, such as of the birds of New Guinea and its many nearby islands, were not necessarily the result of past competition. There was no question that their patterns of distribution did create this impression: two species of certain groups of birds never inhabited the same island. In fact, some combinations of species appeared to be "prohibited." As a result of exclusion through competition, some species might never occur simultaneously.[124] But might this pattern not just as easily be attributed to chance—to a conglomerate of unidentifiable causes?[130]

How could I find out whether the zonation of zooplankton-eaters was the result of interactions between members of the same species or interactions between zooplankton-eaters of different species? How could I find out whether the patterns of furu distribution were the result of past competition? And why had I chosen to work in a lake in which ecologists had to grope around in the dark, while most of the other lakes with cichlid species flocks were so crystal clear? Why was I working with organisms that could go without eating for six weeks, so that it took forever before the consequences of competition for food became evident? And why now, just as our work was getting under way, was the Nile perch—a large predator—swimming into the waters we were investigating? This was the worst possible moment to try to find evidence of competition. The remaining zooplankton-eaters probably had an abundance of food and space, now that the Nile perch was devouring members of their kind so rapidly.

There was definitely structure in the species flock of the furu. But I would probably never be able to find out whether it was competition, other mechanisms, or chance that was the underlying cause. The community was changing at breakneck speed. Might that not be of use to me? There was extensive ecological isolation between the different species of a trophic group. This certainly explained, in part, why so many species could coexist in one trophic group, but were not other forces at play here as well? Each species, even in its own habitat, was only dominant under certain circumstances. The definition of what were optimal circumstances varied from species to species. One moment an environment was optimal for one species, the next moment for another. The population of a particular species rose and fell in accordance with fluctuations in environmental conditions. Mathematical models existed that predicted that instability of conditions in the short term could enhance the stable coexistence of species in the long term.[131] "Stability in the long term" in this case refers to the ability of a biotic community to recover from extreme fluctuations in the density of one species in the system. "Instability in the short term" means that if a trend develops whereby the numbers of a particular species continue to fall drastically, the species will eventually die out. Stability in the long term that can only exist because of instability in the short term. It sounded promising, but I feared that the instability I was witnessing here in the short term—right in front of me in Mwanza Gulf—would lead to instability in the long term. An ecosystem must give, be elastic. But this ecosystem was strained to the limits. One link after another in the food chain was disappearing. The system was under stress. It wouldn't be long now before it burst. I was too late.

8

The Battlefield: Extinction

Emin Pasha Gulf, a Specimen-Collecting Trip, July 1985

It was already light when I awoke. Before getting up, I wanted to lie in bed for a while in my improvised tent. Two chairs—one at the head end, the other at the foot—served as tent poles. A mosquito net hung over the backs of the chairs and was attached with string to the deck of the ferry-boat. I was stiff from a night on the oily, iron deck, but resolved not to complain about it. At least the wanderers had a rubber mat to lie on and a mosquito net. The Africans had nothing with them, except perhaps the gene that offered them protection against malaria. They had to rely on their resilience, which seemed all but infinite.

It was cloudy, with clouds almost as white as the mosquito net. A kite was gliding high overhead. I turned my head so I could follow it as long as possible. It had just disappeared out of sight when three black silhouettes passed by me. Silently they crept toward the empty quarterdeck and stood there. One of them held his arms crossed in front of him, stretched downward. Another stood back to back with the first silhouette and fell to his knees, arms pointed upward as if about to embrace something round and large. The third stood with his legs wide apart, torso twisted and slightly reclined, and arms spread sideways scarecrowlike. The silhouettes remained this way for a moment, motionless, like a group of statues. Then shrill laughter was heard and they began moving. It was Mhoja, Elimo, and Kabika. They began their day dancing, while cleaning their teeth with a wooden stick.

Whoever had been asleep was now awake. When I stuck my head out from under the mosquito net, the dancing stopped, as if by magic. We greeted each other and started laughing, then the Tanzanians resumed dancing. Would the others on this expedition have the same experience? The first events of the morning made a deep impression. In the course of the day receptiveness dwindled, until finally, the senses dulled, one fell asleep in the evening.

A radio was turned on in the wheelhouse. The radio belonged to Katololos, the oldest crew member. Only he was allowed to man the switches. Plaintive music from some North African country alternated with news and political commentary in Swahili. An airplane had been hijacked in Lebanon.

Melle raised the anchor, winding the cable onto a hand winch. It was heavy work. Ill-tempered, he addressed the anchor as if it were a recalcitrant child. Why had it made such a grating noise the night before? He had hardly slept, worried that the ferryboat might become unmoored. As soon as the anchor came aboard, dripping, the boat set off at full steam in search of breakfast. I stood on the forward deck and peered straight ahead through a pair of binoculars. We moved slowly into a small inlet bordered by a seam of papyrus. A derelict wooden boat lay there half beached, its wheelhouse made from the cab of a lorry. In block letters on the dark green door were the words "Ibrahim Transport," and above them, in elaborate handwriting, "*sina makosa*," "no errors made." Behind the boat was a strip of grass on which emaciated cows were grazing. Further inland, the landscape became hilly. On one hill, furrowed like a forehead, was a shed of corrugated iron next to a brick house with white plastered walls. Higher up the hill was a wooded area. The landscape appeared luxuriant, virgin. That such a landscape still existed! Suriname, said one of the wanderers who had grown up there. It was difficult for me to know what I was seeing. The longer I looked, the more the landscape lost its color. The view was transformed into a yellowed black-and-white photograph, an image from a distant past. It probably used to look this lush at many locations along the shore, before the forests were chopped down on a massive scale for making charcoal. The forest was being burned up under maize porridge. The trees were disappearing at an alarming rate, but perhaps these remote islands were not overpopulated. There wasn't a soul in sight.

The boat had reached the middle of the bay. This was as close as we could go to shore. The engine roared flat out for a few seconds before being turned off. Two cackling go-away birds, who seemed to have something to say about this manned colossus, ceased their tropical din. There wasn't a breath of wind. For a moment it was absolutely silent. Then a string of people emerged from the shed as if being squeezed out of a tube. Screaming children extricated themselves from the rest of the group and scurried down to the beach. Near the entrance of the shed, which served as a makeshift church, stood a lone African clergyman in his robes. Arms crossed in front of him, he watched his deserting flock surge toward this unannounced spectacle. Was that the forefathers drifting ashore? What a job it would be to keep the people in church if they were to start landing here again.

Never before had the ferryboat been so far from home. Normally it plowed back and forth across Mwanza Gulf, day after day. For this expedition to the Emin Pasha Gulf and surrounding areas, it had been rented from the German gold-seeker, who was forbidden to search for gold any longer, and fitted up as a laboratory. The half of the boat that bordered on the wheelhouse had been covered with tent canvas, which thus served as a sunshade. There were chairs and long wooden tables, to which microscopes had been attached with carrier straps, as well as tables with planks for measuring, writing materials, and anatomical equipment, such as scissors, scalpels, laryngoscopes, and tweezers. There were also large iron basins for sorting the fish in and dozens of lockable plastic buckets in which to store dead specimens. In the corner lay barrels of formaldehyde and diesel fuel as well as personal luggage.

A man wearing a tuxedo and tattered shorts paddled toward us from shore. Vague geometric paintings were still visible on his canoe. They had been applied in better times, with yellow, blue, and red commercial paint. The man greeted us at length and asked if the trawler they had spotted the day before also belonged to us. The crew answered affirmatively. It was the *Sangara*. With this trawler, the furu were caught that were subsequently examined on the ferryboat. The man in the canoe took our orders, paddled ashore, and returned a short time later with our breakfast. I set the table. Sardines in oil, cooked plantains, bread, and coffee. Like a seventeenth-century still life, only under a blazing sun. A "breakfast piece," bobbing around on the Emin Pasha Gulf.

On the table lay the most recent chart of Lake Victoria, prepared by a certain Captain Whitehouse in 1903. We talked about which route to follow and where samples would be taken during the coming days. At many locations, rocks lurked just beneath the water's surface. Dangerous routes were marked with dotted lines, safe ones with solid lines. I feared the Tanzanians would unsuspectingly follow the dotted lines, and a film reel of accidents passed before my mind's eye. Most of the crew couldn't swim and the insurance wouldn't cover anything if we followed the dotted lines.

As soon as Katololos realized that the prescient approach was playing tricks on the Europeans, he began complaining aloud: "These routes are fine. *Hamna wasi wasi*, no problem. Wanderers are pessimists. Always fussing about dangers. No wonder you laugh so little."

He paced back and forth in front of our breakfast table, looking irate.

"They don't even love children," he said to one crew member. "They have two, or one. Even one is too much for some. Dying childless but with an enormous pile of insurance policies in the cupboard. As if it wasn't all in God's hands."

He shook his head.

"And don't tell me it isn't true. I was trained as a master fisherman in Holland. I used to eat your raw herring."

He threw back his head, opened his mouth wide, and held an imaginary herring by the tail above it between his thumb and index finger. Two texts came to mind that I had once seen painted on the walls of buildings in the heart of Amsterdam. One had read: "People die and are unhappy." The other, about a ten-minute bicycle ride from the first, offered the following advice: "Live, boy." Katololos could have written both.

There was still time to negotiate with Katololos, because for the time being we could go no further. We had run out of ice. Ice was essential, because the furu died peacefully and quickly on it while retaining their shape and color. A furu that has died on ice is suitable for inclusion in a museum collection. The ice had disappeared very quickly and a new supply was needed. But there wasn't an ice machine to be found within a radius of several hundreds of kilometers. The *Sangara* had been sailing all over the Emin Pasha Gulf for days now in search of an operative radio link with Mwanza: "Who will bring us ice, a cubic meter of ice?" There had been no reply, only deafening static.

Close to the ferryboat a fish eagle swooped down and dug its talons into a Nile perch that had been thrown overboard by a crew member.

Body bent, the Nile perch was floating on the water, its mouth open: its swim bladder had popped out of its mouth because of the difference in pressure between the lake bottom and surface air. A balloon-blowing corpse. The eagle tried to fly away with its prey but didn't succeed immediately. It dragged the heavy perch for meters, scarcely increasing in altitude. Finally, it gained speed and managed to fly off to its giant nest. The Nile perch had been caught by the *Sangara* in deeper waters. It may have been the first Nile perch the fish eagle had ever seen.

There were very few Nile perch (a species of the genus *Lates*) in Emin Pasha Gulf, and they were certainly not to be found in shallow inlets such as this one. But it was only a question of time before this introduced predatory fish entered this particular area. Like concentric rings around a stone that has been thrown into the water, the Nile perch appeared to be spreading, wavelike, from the point in Uganda where it had been introduced. It was an enormous predatory fish, eating its way through the lake like a giant vacuum cleaner. That same day, the *Sangara* returned to the ferryboat. Ice was on its way and would reach us shortly. The biologists on board felt they should continue working feverishly. A massive front of millions of Nile perch was breathing down our necks and everyone was acutely aware that this was the last opportunity for observing the original fauna.

After the arrival of the all-essential ice, we spent several weeks collecting as many data as possible about patterns of distribution and geographic variations in the form and color characteristics of known species. New species, too, continued to appear. They were given temporary names and preserved in formaldehyde. We felt like writers of epitaphs: in this jar lies Red-tailed Dented Head, in life, an avid prawn-eater. It was a depressing occupation. Fully aware that most of these species were doomed, we were catching and killing them. The argument for doing so was that we wanted to save them from anonymity. We wanted to give them names, make drawings and pictures of them, and describe as many of their biological features as possible—to ensure that their massive disappearance would not go unnoticed. The least we could do was to register what was happening and publicize it.

A Man with a Bucket

I have frequently referred to the Nile perch and its voracious appetite. I have also mentioned the fact that it was not indigenous to Lake Victoria. But I have not yet described how and when the Nile perch came into the lake. For a while I asked everybody whom I thought might know. An old woman who had spent her entire life on the shores of Lake Victoria said: "*Wazungu mbwana*, wanderers, sir," to which she added: "The white people put the baby monsters in the lake to help us." She burst out laughing, bent forward, and placed a shopping basket, out of which protruded a bouquet of dazed chickens, on her head. As she waddled away, without looking back, she said: "See you, wanderer, and thank you."

During the 1950s, a number of British colonial officials who had been hired to improve the Ugandan fisheries had considered introducing a large predatory fish in Lake Victoria. The idea was that it would feed on the predominantly small, bony furu, which weren't very popular among the Ugandans.[132] The Nile perch, an alien to the lake, was mentioned as a possible candidate. It was a giant fish suitable for eating. Geoffrey Fryer, an ecologist well versed in the East African lakes, fiercely opposed the plan. In 1960 he published an article in which he discussed at length the disastrous consequences that an eventual introduction might have.[133] Extremely little was known about the lake's flora and fauna. Most of the species of fish had neither African nor scientific names and the inventory of plankton, insects, and other invertebrate organisms was far from complete. Scarcely anything was known about the structure of the food web. It would be very premature to start playing around with the ecosystem at this stage, particularly someone else's ecosystem. Manipulation of an almost undisturbed ecosystem is never recommended—least of all when so little is known about it. Moreover, of the different forms of manipulation, the introduction of alien species is—because of its irreversibility—one of the riskiest. No one could anticipate the extent of the risks if the Nile perch were to be introduced. For most biologists and conservationists, this was reason enough to reject the idea; for fish technologists it was precisely the reason to take a gamble. Despite Fryer's warnings, in May 1962 Nile perch from Lake Albert were introduced near Entebbe in Uganda. In 1963, several Nile perch originating from Lake Turkana were also introduced at Kisumu in Kenya.[134] One of the arguments for releasing the perch in the lake was striking: they were already there. A breeding population had been found in the lake. But how had they got there in the first place? Had they escaped from hatcheries or had someone thrown them in at his own initiative? Every time I thought about it, I was amazed that for the total disruption of the largest tropical lake in the world, nothing more had been needed than a man with a bucket.

In Bruce Kinloch's *The Shamba Raiders: Memories of a Game Warden* from 1972, the following passage appears about the introduction of the Nile perch in the small Ugandan Lake Kyoga, not far from Lake Victoria:

> In September and October 1955, J. S. caught several hundred small Nile perch in a seine-net in Butiaba Bay on Lake Albert and successfully transported them across Uganda ... to release them in the upper Nile below Owen Falls Dam and ... Lake Kyoga, which is little more than a shallow valley drowned by the Nile. Further similar stockings were carried out in 1956 and 1957, in which D. R., A. A. and H. S. of the Fisheries Division of the Game and Fisheries Department also took part [made anonymous, T. G.]. The results of these operations were, to say the least of it, spectacular! ... in fourteen years, largely as a result of stocking with Nile perch, the commercial fish production from Lake Kyoga rose from 4,500 tons to nearly 49,000 tons—an increase of over 1,000%! And of this prodigious total the Nile perch now comprise more than 56% of the overall catch—a case of what the Americans call "trash" fish (the smaller and less palatable fish which form the bulk of the Nile perch's food) into a ready catchable and marketable product.[135]

This passage[136,137,138] illustrates the way many thought about introductions at the time. In poor countries such as Uganda the enthusiasm about heavily increased fish production was certainly understandable. But the possibility that this high production might only last for a short while appeared to be considered unimportant in these circles.

In July 1986 a letter was published in the "Letters to the Editor" section of the Kenyan newspaper *The Standard*, signed by J. Ofula Amaras. Amaras, a former Kenyan fisheries officer, reports having single-handedly introduced the Nile perch into Lake Victoria:

> I have been following with keen interest the argument raised by experts about the existence of Nile perch in Lake Victoria. What I have observed is that most of the experts do not know the year the Nile perch were stocked in Lake Victoria and where they were transferred from or the purpose why perch were stocked in Lake Victoria.

... Nile perch were stocked by me from the then Lake Albert (Uganda) in August 1954 after getting clearance from former Senior Fisheries Officer of Uganda Government by the name Mr. A. A. [made anonymous, T. G.].

Therefore it should be in record that the perch were stocked in Lake Victoria in 1954 not in [the] 1960s.

Amaras justified his actions as follows:

As to whether the Nile perch is causing hazards to other species of fish in Lake Victoria is for experts to prove. What I can prove to them is that Nile perch has never caused any hazards to other species in the Lakes, Albert in Uganda and Turkana.

That the indigenous fish of Lakes Albert and Turkana were able to coexist with the Nile perch did not mean that the same would be applicable in Lake Victoria. The fish still found in Lakes Albert and Turkana had either undergone adaptations that allowed them to resist predation by the Nile perch or occupied habitats where the Nile perch preferred not to go. Species originating from the same habitat as the Nile perch had had the time, evolutionarily speaking, to find a way of dealing with the presence of these predatory fish. Most of the species of furu in Lake Victoria had not been granted this time. But before describing the loss of a large number of these species, I would like to leave Lake Victoria for a moment.

The Magazine *Extinction*

The extinction of species is not an unusual phenomenon. Since life began some 3.5 thousand million years ago, it is estimated that more than 99 percent of all species that ever lived on earth have died out. In my search for theories about extinction it surprised me that, compared with other evolutionary-biological subjects, so little could be found. There were complete libraries about the origin of species, but I would be surprised if a single shelf could be filled with papers on extinction. I searched in vain for the magazine *Extinction*. It didn't exist. Might biologists be ignoring this depressing subject? In any event something unusual was happening, because if almost 100 percent of all the species that ever existed have disappeared, then extinction is a normal phenomenon, not much less common than the origin of species.[123] Raup, a paleontologist from the University of Chicago, has likened the neglect of the subject "extinction" to a demographer who studies the growth of populations and, in doing so, pays close attention to births while ignoring deaths.[139]

Traditionally, paleontologists have always been preoccupied with the extinction of plants and animals but have paid little attention to the causes of extinction. This is understandable because it is seldom possible to identify, with any certainty, the causes of the extinction of species, genera, families, and even higher taxonomic groups from the geological past. Even when many good fossils are available this cannot be done. It is possible to document the geological era during which certain fossils appeared in deposits and when they disappeared. The numbers of individuals of the various species that are present can be determined and links sought with changes in the environment. A hypothesis about the background of the disappearance of a species gains credibility if several correlations point in the same direction, but experimental proof allowing one to speak of a causal connection and not of a coincidence cannot be furnished.

In contrast to most paleontologists, ecologists have always been interested in the mechanisms responsible for the dynamics of populations, but they focus mainly on present-day organisms. How do species alive today manage to survive in an ecosystem? This is the central question. Nowadays, the answer is, with increasing frequency: poorly. Species are disappearing at an alarming rate.[140] Sometimes a species dies a taxonomic death but is survived by a few offspring. This is a form of pseudoextinction that I am not concerned with here: the species evolves until at a certain point in time the changes in features become so radical that taxonomists cut the knot and give the changed form the status of a new species. Today, species really are dying out, that is, leaving no offspring—like a dead-end road.

Biologists are loath to claim that a species has become extinct. It can seldom be said with certainty that the last potential pair has disappeared. And there is always the hope that somewhere, hidden away, unnoticed, several individuals are still thriving.[123]

The extinction of all individuals of a species is definitive, but the disappearance of species over a large part of their area of distribution can also represent a major loss. When entire populations are wiped out, an important part of genetic variation—and evolutionary resilience depends on this variation—disappears. Lack of genetic variation makes species vulnerable and increases their chances of extinction, because no individuals exist with the necessary genetic equipment to survive under changed circumstances. Conservationists continually wrestle with the problem of the minimum number of individuals that need to be protected to ensure the survival of a species. Individuals need to be able to find a mate, the genetic variation must be great enough to facilitate reacting to slow changes in the environment, it must be possible to meet chance adversity, and so forth. In the best-known version of the legend of Noah's ark, one pair of each species entered the boat. That is usually too small a basis for a population to be viable. But studies of original Bible texts have revealed that seven pairs of each species boarded the ark. According to Foose, an expert in the field, seven pairs of one species can carry a considerable portion of the genetic variation of the population. That is, if the animals are not relatives.[141] With a view to this conservation problem, MacArthur and Wilson introduced the concept of "minimal viable population." A population with more individuals than this critical value runs no direct risk of extinction. Depending on the reproductive capacities of the organisms, the "minimal viable population" can vary from several dozens to several hundreds of individuals.

Quite distinct from the disappearance of a species is the disappearance of an entire genus, family, or order. Sometimes several families, orders, or even higher taxonomic groups disappear en masse over a relatively short period of time.

The key question in the search for the origins of extinction is whether victims fall randomly or not. Do groups of organisms die out because of unfortunate circumstances or because, compared with other organisms living in the same area, their genetic equipment is inferior? Causes of extinction that have long been acknowledged are climatological changes, fluctuations in sea level, predation, epidemic illnesses, and competition with other species. Raup rightfully remarked that this list is not free of anthropocentricity: these are precisely the factors that threaten man in his daily existence and with which he is thus familiar. Presently, attention is also being given to causes of extinction that were less prevalent in the past but that, unfortunately, have become common today: chemical pollution of the oceans, changes in the chemistry of the air, and cosmic radiation.[139] Great interest is also being shown in the effects that cometary impact may have had on the extinction of organisms.

Periods or events during which the majority (more than half) of all species on earth disappear are extremely scarce. They are believed to occur approximately once every one hundred million years. In comparison, the death of a species is commonplace: a species exists on average for only one million years.

Paleontologists have documented only five periods in the fossil archive during which most of the biological diversity disappeared within a relatively short period.[139] The best documented of these periods is the one marking the transition from the Cretaceous to the Tertiary, approximately 65 million years ago. Since much has been written about this mass extinction, I will limit myself here to a summary, to illustrate how a paleontological investigation into the causes of extinction is carried out.[142] Of necessity, the paleontological approach differs considerably from the study of present-day cases of extinction, such as our work on the furu.

The end of the Cretaceous period also signified the end of the golden age of the reptiles, during which many forms of dinosaurs had dominated life on land for more than one hundred million years. These animals are believed to have disappeared over a relatively short period, ranging from several hundred thousand to several millions of years. During the past few years, though, there have been a growing number of indications suggesting that it may have gone much faster. Not only dinosaurs disappeared during this transition period between the Cretaceous and the Tertiary; so did many species of organisms and vegetarian plankton in the sea. Many species became extinct during a short period lasting ten to one hundred years. It may seem incredible that a period from such a distant past can be established with such accuracy, but a growing number of paleontologists appreciate the merits of this hypothesis. In 1980, Alvarez and colleagues posited that during the late Cretaceous-Tertiary, the earth was hit by a comet measuring approximately ten kilometers in diameter. The amount of energy released by such an impact is enormous. For several years, the earth is believed to have been shrouded in a cloud of dust through which sunlight could not penetrate, so that photosynthesis became impossible for land and marine plants. The vegetarians were the first to disappear, followed quickly by the carnivores, who had lost their main source of food in the vegetarians.

In support of this hypothesis, it was suggested that during the same transition period a layer of iridium settled over the earth. Iridium is a metal seldom found in the earth's crust but occurring frequently in comets. Critics of the impact hypothesis pointed out that the iridium could also have originated from the earth's mantle. Based on the arguments presented up till then, it was not possible to identify an unequivocal cause—earthly or extraterrestrial. But the debate didn't end there. The advocates of the impact hypothesis predicted the existence of a gigantic impact crater, which indeed was found several years ago in the Yucatán peninsula in Mexico. It has a diameter of 210 kilometers and is about the right age. The fact that a large comet did collide with earth has therefore now been established. Yet even if the collision took place at the "right" moment, the possibility that other factors may also have played a role in the mass extinction of all those groups of organisms cannot be excluded, particularly because the late Cretaceous-Tertiary is known to have been one during which major shifts and faults developed in the earth's crust.

Beginning in 1984, a series of articles by Raup and Sepkoski were published that caused quite a stir.[143,144] They analyzed a large quantity of data from the literature on fossilized marine organisms. The last mention of a taxonomic family in the fossil record was taken as the moment of extinction. The disappearance of families of these organisms was found to take place not at random intervals but cyclically. Raup and Sepkoski identified twelve periods over the past 250 million years during which a striking number of families became extinct. Nine of these twelve periods occurred at intervals of 26 million years. In addition to the comet-impact hypothesis (the comet is supposed to have resulted in the iridium deposit), there was now a hypothesis about the periodicity of the extinction of marine organisms. Once every 26 million years, a comet or a shower of comets was believed to have collided with the earth. The idea was an attractive one. Ecological communities are so complex that ecologists sometimes wonder if anything will ever be understood about them. A giant collision at regular intervals might give them something to go by; if evolution is radically interrupted every 26 million years, then ecological mechanisms, as an explanation for patterns of evolution, would become secondary. Several researchers focused immediately on the periodicity hypothesis. Criticism was leveled at the statistical analyses and the effects of the time scale chosen to establish the periodicity, and alternative hypotheses were put forward that explained the appearance of the iridium layer without reference to extraterrestrial forces.[145] Any study seeking to identify cycles in the origin and extinction of organisms is doomed if the taxonomic units are incorrect. And it is precisely in this area that Raup and Sepkoski's work revealed weaknesses. As a result, the periodicity hypothesis is a tenuous one.

The paleontological study of the extinction of organisms inevitably has a speculative aspect to it. This deficiency can largely be compensated by studying the dynamics of species that still exist today but are threatened with extinction. Unfortunately, this approach has its own disadvantages, as most of the species now on the verge of extinction are threatened in their existence by man. But who can say with certainty that the causes of anthropogenic extinctions are the same as the forces responsible for disasters in periods before man existed?

Vulnerable Furu

Between 1977 and 1983, we biologists from Leiden occasionally caught a large Nile perch in Mwanza Gulf, but none of us saw anything alarming in this. Fryer's article from 1960 had long been forgotten and only in the form of unsubstantiated reports did we learn that the number of furu in the Kenyan Winam Gulf was declining rapidly.

I don't believe it ever occurred to me that the Nile perch might represent a threat to the furu species flock, even though it was appearing with increasing frequency on our dinner plates. That is, until April 1985, when it started to dawn on me. Mhoja, Elimo, and I were bobbing around in the wooden boat somewhere in the middle of Station G in Mwanza Gulf, the larvae-ridden boat a sieve, Elimo forever bailing, Mhoja filling the holes with cotton. But their efforts were inadequate. Slowly but surely we were sinking. What did our field biologist do? He stuck his nose into a putrid bucket. An unusually small catch and an unusual combination of species. Species were missing that had never been missing before, not once in all those years. And fish from deeper waters appeared in numbers he had never seen before. A feeling of bewilderment, of disorientation, overwhelmed him. A vague sense that the Nile perch might be behind this. The image of an army of Nile perch massacring the indigenous fauna.

Fanatically concentrated as I was on the question of how species of furu originated and coexisted, it took, in retrospect, an amazingly long time before I realized that species were actually disappearing. And when I did realize this, I thought it sad but, strangely enough, uninteresting. I was busy creating speciation models and searching on a micro-scale for indications of the effects of natural and sexual selection. I wanted to establish how the structure of this species flock had evolved. I was searching relentlessly for indications of the existence of competition in terms of space or food. But I had no eye for the fact that species were disappearing. It was not until my fish were nowhere to be found except in the stomachs of the Nile perch that I decided the time had come to include this predatory fish in my research program. The stomachs of the Nile perch were bursting with furu that—often still whole and wearing surprised expressions—had begun decomposing. Until recently, biologists were not prepared for the fact that they might become witness to the extinction of species. But this changed overnight. Today, young researchers at conferences speak of poisoning, pollution, and the large-scale extinction of species without batting an eyelid, as if it had never been otherwise. Extinction has become commonplace and the destruction of ecosystems the norm. The general public has been bombarded with ecological disasters, with the result that the disappearance of life on earth, after being in the limelight for several years, no longer holds their interest.

Imagine that our knowledge of the furu had only been derived from a rich fossil archive. Of the more than three hundred species—each of which is probably a combination of species—at most we would have been able to distinguish fifty forms on the basis of the fossilized remains. This is because the most important features used for identifying species—the colors of the sexually active males—do not fossilize. Only later did I realize how advantageous my position was from this point of view, compared with that of the paleontologists. For the time being, though, I didn't even want to accept the fact that species were disappearing at an increasing rate, let alone that I felt pleased about being able to study the dynamics of extinction firsthand. The enforced metamorphosis from ecologist to paleontologist of the youngest fossils on earth did not come easily. Meanwhile, the Nile perch went on filling its stomach and producing unlikely quantities of progeny.

An explosive increase in the Nile perch had already been observed in Winam Gulf in Kenya in 1978–1979.[146] (These data were only published in 1983.) At the same time, the furu had disappeared from the catches. The first sizable catches of Nile perch in Mwanza Gulf were not made until several years later, during the last months of 1983. Here, too, the number of furu had declined rapidly. As I have already mentioned, Nile perch were scarcely found at all in the western part of the lake at the time. But less than three years later, that picture had changed completely. A catch that had previously consisted primarily of furu now included only a few of these fish—crushed under a mountain of Nile perch.

Contradictory reports on the effects of the Nile perch appeared frequently in scientific journals as well as in daily and weekly newspapers. Some biologists believed that the Nile perch could only be held partially responsible for the disappearance of the furu. A more important reason, in their eyes, was overfishing.

It was true that the number of furu in Mwanza Gulf had already declined before the arrival of the Nile perch owing to overfishing, and perhaps this had hastened the effects of the Nile perch. But nowhere in the lake had it been established that furu communities had disappeared owing to overfishing alone. Thinly populated areas infested with tsetse flies offered an opportunity for testing the theory, because very little fishing took place there. But there too the furu had disappeared at a rapid pace as soon as the Nile perch had arrived.

There were also laymen and even biologists who maintained that nothing was wrong. According to them, furu could still be caught in abundance. But none of their articles contained data on the number of species present before the arrival of the Nile perch, let alone proved that the numbers of individuals had not declined.

Details were known only about Mwanza Gulf and the immediately surrounding areas. Our group took samples from a series of stations right across Mwanza Gulf during the period 1979 to 1990. In this area and on the nearby rocky islands a total of 123 species were found, most of them undescribed. In the Dutch National History Museum at Leiden, there were still cellars full of furu from Lake Victoria, waiting for dedicated, unambitious biologists prepared to devote their lives to describing organisms that had long since ceased to exist. The picture that emerged

Figure 8.1
Catches made by trawlers in Mwanza Gulf. Before (left) and after (right) the proliferation of the Nile perch.

once all the data had been examined could be rendered in a single word: a battlefield. Since we had started studying the ecosystem, the number of species had only declined. The same applied to the number of individuals of the not yet completely obliterated species.

A large number of furu species from Lake Victoria were extremely vulnerable. They were limited in their distribution to this lake and, what is more, to only a small part of it. Some species occurred in only a single inlet of Mwanza Gulf. Organisms with such a limited distribution are a priori more likely to become extinct than species with a wide distribution.

Broadly speaking, Raup, who posed the question of to what extent selectivity played a major role in the extinction of species, distinguished three models:[139]

Figure 8.2
Catch during the transition from the era of the furu to that of the Nile perch.

1. The extinction of species occurs randomly: in other words, "fitness" differences between individuals of different species play no part. Raup used the following image: all individuals inhabit a field where bullets are flying in all directions. The extinction or survival of a species is a question of luck.

2. Extinction is the result of a "fair game." Species die out selectively, in the Darwinian sense. Individuals of certain species are better adapted than individuals of other species. The latter species disappear.

3. Extinction occurs wantonly yet selectively. However, this selectivity is not the result of the relatively poor "fitness" of the individuals of the extinct groups of organisms. Surviving species are not better adapted to their normal environment than disappearing species. Raup used the following example as an illustration: insects are much better equipped than mammals to cope with a high dose of radioactivity such as might be released during a nuclear war. Compared with insects, many species of mammals would die as the result of such a war. This is an example of selective extinction. But insects' improved chances of survival are not the result of natural selection favoring organisms equipped to resist radioactivity. Their immunity is therefore not an adaptation but an evolutionary by-product.

I shall attempt to interpret the extinction of species of furu on the basis of Raup's ideas. But first the facts.[147] It was striking that in the open waters the fish-eaters had disappeared first. Only after that had happened did other groups, such as the snail-eaters, insect-eaters, and mud-biters, disappear. The last surviving group was the zooplankton-eaters, although they, too, were heavily affected. In 1987–1988, three furu were caught—all of them zooplankton-eaters—at Station G, the place from which most of the samples had been taken. Before the arrival of the Nile perch, the catches would have yielded at least nine thousand furu. In the open waters, all species except three (93 percent) had disappeared. More species had survived at shallower locations (30 percent), although there was also a sharp decline in their numbers. It was primarily the snail-eaters, insect-eaters, and epiphytic algae-grazers from the littoral zones that had survived. Initially, at least thirty-one species were found along the stretches of rocky coast and near the rocky islands in the vicinity of the area from which the samples were taken. Eleven of these species were restricted mainly to the rocks. The rocks were an important, but not the only, habitat for another eight species. Lastly, there were several species that only occasionally occupied rocky habitats.

Of these occasional visitors, not a single species was left in 1990. Half the species that had regularly inhabited the rocky stretches had also disappeared and several of those species that had inhabited only the rocky areas had vanished. In total, more than 80 of the 123 species (approximately 70 percent) had disappeared from the sampled area and the adjacent rocky habitats.

An Interpretation

Had those species that could no longer be found in the sampled area really become extinct? This area covered only a small part of Mwanza Gulf and the gulf represented only a small part of the whole lake. Unfortunately, the same picture emerged from each of the many locations in Mwanza Gulf where random samples were taken. Moreover, samples taken outside the gulf strengthened our conviction that extrapolation of the data was justified. This would mean that approximately two hundred of the more than three hundred species were threatened with extinction or had already become extinct. Never before in recorded history had the mass extinction of vertebrate organisms occurred on such a scale or within such a time frame.

Was the survival or extinction of furu species selective or random? And if it was selective, were the surviving species better adapted to deal with the confrontation with the Nile perch?

Even though we were so close to reality and had access to much more detailed data than paleontologists, interpretation remained difficult. It was clear that the metaphor of a field in which bullets were flying around at random was not applicable. The Nile perch were not randomly distributed over the lake. They resided primarily in the deepest regions of the open water, occurring much less frequently in the littoral zones and near rocks. This explained why fewer species had disappeared in those areas. The same applied to the species in the open water. There, the only surviving species were the zooplankton-eaters. The habitat of these pelagic species only partially overlapped with that of the Nile perch. It was striking to note that the snail- and insect-eating species, which lived in the littoral zones or near rocks and which never left their habitats, had survived. But all the species that at a certain moment in their life's cycle had left the littoral zones for open water had been eradicated.

That the surviving species were found in the littoral zones and were restricted to rocks or higher layers of water had nothing to do with the Nile perch. This had already been the case before the first Nile perch made its way into Mwanza Gulf. In other words, the strong selection pressure exerted by the Nile perch could not be held responsible for the current spatial distribution of the surviving species. These species survived the onslaught of the Nile perch not because they fared better in the confrontation with this predatory fish (the "fair-game" model), but because of other factors, unrelated to the Nile perch, that were once crucial in determining their habitat. The third model mentioned above—that of wanton extinction—is probably the model most applicable to the furu.

The species from the open waters had disappeared within a few years—that is, within several furu generations. Assuming that the evolutionary potential for survival was present in the first place, this period of time was too short for many species to develop a strategy that would allow them to coexist with the Nile perch. The development of defense or

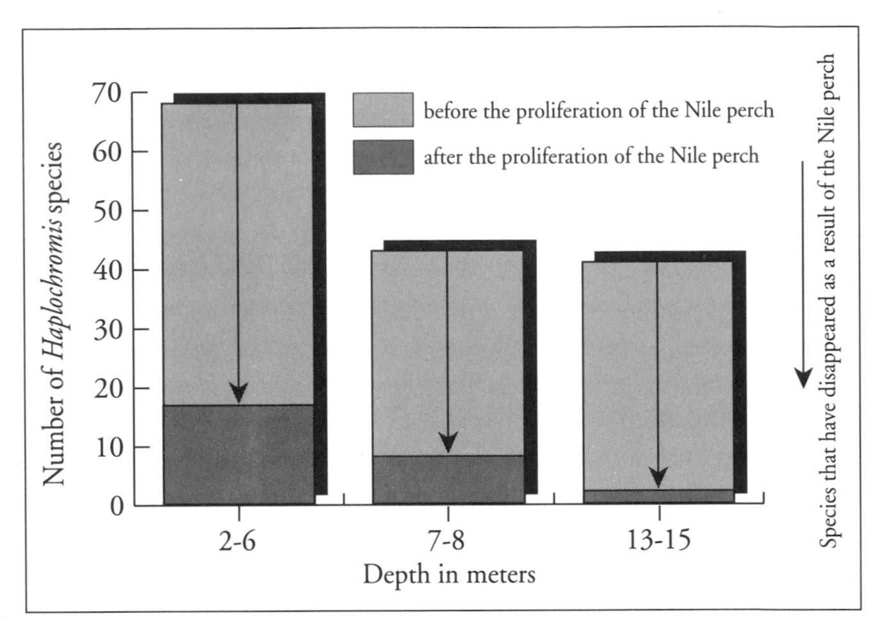

Figure 8.3
Reduction in the number of species of furu in Mwanza Gulf.

escape mechanisms or a shift to another reproductive strategy by means of an evolutionary response is not the only way of responding to a confrontation with a predator. Sometimes natural selection produces organisms with a "plastic" appearance, which can change according to conditions. The external appearance of butterflies of the genus *Bicyclus* changes in successive generations according to whether they emerge during the wet or dry seasons. They have deterrent eye-spots in the wet season but take on a camouflaged appearance in the dry season. Identical genetic equipment that, depending on the temperature to which the organism is exposed during its early development, produces two totally different phenotypes.

But sometimes an adequate response to the arrival of a predator can be found *within* the lifetime of an individual. For example, a change of body form that offers protection against the pike has been found among Crucian carp (*Carassius carassius*).[148] Carp populations in lakes with pike (*Esox lucius*) consist of a small number of relatively large individuals. When predatory fish like the pike are absent, carp populations consist of large numbers of small individuals. In order to study this phenomenon more closely, carp were kept in ponds. Pike were released into half of the ponds. After twelve weeks, there was a difference between the body forms of the carp in the ponds with pike and those without pike. The distance between the back and underside had increased: they had become deeper-bodied. This could be a consequence of selective predation, in other words, of the pike eating primarily the relatively slender carp, with the result that deep-bodied individuals survived. But as it turned out, this was not the case. The body form of the individual carp in the ponds stocked with pike really *had* changed: it was a phenotypical response to the presence of predatory fish. Deep-bodied carp are difficult or impossible for pike to eat. The latter prefer specimens that are more slender than they are actually capable of consuming. The bodies of the carp in the ponds with predators had become just deep enough to prevent them from becoming a potential prey for the pike. If relatively deep-bodied carp are in less danger of being caught by predators, why do slender carp continue to exist? Probably because deep-bodiedness is not advantageous under all circumstances. Deep-bodied carp offer relatively more resistance to water, so that swimming becomes "costly" in terms of energy. If no predators are present, there is no reason for going to this expense, and when competition for food is heavy, such economizing can be of vital importance. It is logical that pike occur more readily in some areas (or in some periods) than in others. But less apparent is how the change in form comes about. Some carplike fish react to alarm substances expelled by members of their own species after they have been seized by a predator. The trigger could thus be a chemical one.

After the explosive increase in the Nile perch population, several species of furu also changed form, like the carp. But we do not know whether these form changes were phenotypical or genetic. Unlike the carp, some species of furu became slimmer rather than deeper-bodied. Perhaps this enabled them to swim more quickly, thus giving them a greater opportunity to escape.

The selection pressure of the Nile perch was too strong for most of the species of furu in the open water to respond to in evolutionary terms. Yet several species did manage to survive in the habitat of the Nile perch. During the next few years we would probably learn to what extent their distribution and life cycles had changed as a result of the confrontation with the Nile perch.

When species become extinct, with the subsequent decrease in biological diversity, this creates ecological and geographical space for renewal. Creation of space is the primary role of extinction in evolution. True form renewal is often only possible after old forms have perished. The fossil archive reveals that many new forms commonly appear following the mass extinction of old species. The best-known example of this is the rapid acceleration of the radiation of mammals that occurred following the extinction of the dinosaurs.[139]

Perhaps the disappearance of most of the species of furu in Lake Victoria would create space for secondary radiations of surviving species or of totally different organisms that were better equipped to deal with the Nile perch. It was not possible to predict how much time this would require.

With respect to biotic communities on coral reefs, we believe it can take millions of years for a new community of coral reef inhabitants to evolve, following the mass extinction of an old reef community. But, compared with most vertebrate organisms, furu-dominated fish communities develop extremely quickly. The surviving furu are even changing so rapidly that a biologist can witness some of the changes during his working career.

The most depressing aspect of the mass extinction of the furu species flock was, of course, that it was a common phenomenon. Wilson estimated the number of rain forest-dwelling species that become extinct annually. His most optimistic estimate was 27,000 species a year. This works out to 74 species a day, or three an hour.

If it is true that, before the human factor began to play a role, the life expectancy of a species was approximately one million years, then about one in every million species disappears annually.[123] Commenting on this, Wilson remarked that the chance of extinction among species in the rain forests has become one thousand to ten thousand times greater as a result of human activity. He concludes: "Clearly we are in the midst of one of the great extinction spasms of geological history."

What can we still do? A growing number of biologists, limnologists, and anthropologists have become actively involved in trying to conserve Lake

Victoria and its biological diversity. They are united in the Lake Victoria Research and Conservation Team, which includes our group. The furu have since been included in the book of endangered species of the International Union for the Conservation of Nature (IUCN). A Captive Breeding Program has also been launched. Some forty species have been sent to Europe and the United States. Each fish was packed in a plastic bag filled with water and an oxygen bubble. Thanks to the mission airplane, which was normally only used to transport the sick, cheese, and gin, it was possible to transport the fish to Dar-es-Salaam and from there on scheduled flights to Europe. The fish have now been housed in numerous zoos throughout Europe and the United States and are continuing to be bred there.[149] Some biologists hope that one day it will again be possible to release several species of furu in Lake Victoria or in small lakes in the vicinity. In any case, it is possible to study the behavior of these fish as long as they are in captivity. There are also plans to create fish reserves in which some of the species of furu can be protected from the Nile perch. But the most pressing issue is to study the changes in the lake as closely as possible, now that the ecosystem is permanently under stress.

9

The Savior: An Ecosystem under Stress

Return to Mwanza, 1989

Levocatus picked me up at the address where I was staying in Mwanza. He appeared not to have aged at all in the three years since I'd left. We grasped each other's hands warmly. As I was about to withdraw mine, European-fashion, I realized just on time that the extended hand should only be released after a repeated squeezing and massaging of each other's thumbs. Meanwhile, I wanted to tell Levocatus that I'd missed him, and quickly sought the word for "missing" in Swahili. To no avail. "I was unhappy not to see you then and I am happy to see you now," I stammered, irritated. We walked toward town and passed the little station dating from colonial days that lay half hidden behind a clump of trees. Something had changed, but it took a moment before I realized what. The plaster walls had been painted and the pointed, tiled roof repaired. The station made a less weary impression.

The economy, which had progressively deteriorated during the early 1980s, had recovered somewhat. Nyerere had adhered to his rigid socialist policies—based on "self-reliance"—until the bitter end. The Tanzanians had suffered under the stubborn integrity of their president, particularly during the last years of his administration. He hadn't even wanted to give the impression that he was making concessions to the West. The importation of foreign goods—which only cultivated dependence—was limited to an absolute minimum. The result was that less and less became available, since the country itself produced very little. But Mwinyi, the new president, was a less dogmatic socialist: he was less averse to foreign investors or the importation of foreign goods. Sure enough, before long, pots of paint and new roof tiles began to appear. Private initiatives were given greater leeway, and involvement in trade was no longer discouraged: essentials, such as soap, paraffin, matches, and medicine, became available again and even refrigerators, video recorders, bicycles, and motorcycles were sold.

"Just look at the Indians," said Levocatus. He stopped and took hold of my shoulder. "They have everything in their shops, absolutely everything. *Mpaka gari*, even cars." He paused for a moment, before adding: "For those who can afford them." I could hear from the resigned tone of his voice that he didn't consider himself a member of that group.

Cheerful music emanated from a loudspeaker. I caught a few fragments of the Swahili text: "Of course you want to eat well ... oh, darling, if only I could get my hands on that medicine called money ... to kill someone. That's going a bit far ... God ... forbidden ... tralala."

The square in front of the station was filled with Nile perch packed in beautiful structures made of bent branches and grass. These "bales" looked very normal, as if they had been made that way for thousands of years. Yet I was seeing them for the first time. It surprised me that during such a short time an entire industry had emerged. Filleting factories and smokehouses had appeared, and Nile perch were also being canned. Perch were being exported on a grand scale via Kenya to Israel and Europe. Melle had anticipated well during our time here. Not long ago, in an Indonesian restaurant in Amsterdam, a Sulawesian fish platter was recommended to me. I was eager to try it. After it was served, I began, unsuspectingly, to enjoy it, fancying myself in Asia—until the vague taste of mud supplanted the smell of coriander and threw me with a jolt back onto the ferryboat heading for the Emin Pasha Gulf. It reminded me of how, for weeks on end, we had lived off a diet of Nile perch with rice and tiny stones, while trying to overtake the army of Nile perch—in the hope of finding yet unspoiled areas. This was unmistakably Nile perch clothed in a spicy Asian jacket.

In a covered gallery facing some offices and a luggage depot, a crowd was sitting on the ground waiting for the train to Dar-es-Salaam. It was quite possible that the *treni* would start moving some time that evening, and once it started moving, there was no stopping it. Those who hadn't reserved a ticket weeks in advance arrived early in the hope of securing an unclaimed seat. The waiting crowd had all the time in the world: packed together under the roof of the narrow gallery, that mass of time was almost palpable. I caught myself wanting to claim some of it. But wouldn't I prefer to have really arrived here—to have shed the Western notion of time? That is, if there is any way back for a Westerner who has allowed himself to be dragged along the narrow path of congested diaries toward his final destination?

It was early—the third hour of the day had not yet passed—and it was already hot. A soft, pleasing wind was blowing. "*Mvua*, rain," said Levocatus, pointing to the sky. A small, compact cloud hung above us like a stain on the clear blue backdrop. A final convulsion of the rainy season. The drops didn't even reach the ground any more; steaming, they shied away from the earth.

We walked from Mwanza along the rails toward Nyegezi, past industrial sites, shipyards, and sawmills. On the walls of derelict sheds were crude drawings of sexual encounters. I was grateful I wasn't a Tanzanian woman.

A group of boys were seated next to a gas station under a tree, surrounded by piles of car tires. Tire jacks, a small foot pump, and a box with supplies for patching a tire formed the mainstay of their operation. One boy, who had been afflicted by polio, scrambled across the street as quickly as he could and nestled against the tire-patchers' tree. His knees were scraped raw. Why didn't he protect them with a piece of tire?

Levocatus saw me watching them and said: "His family feeds him." Then he asked me about my family. "How are your parents? And your wife? Has she given birth?" He shook his head when I told him I was still childless, adding: "And you didn't pay anything for her if I remember correctly?"

Piles of charcoal-filled burlap sacks lay stacked high along the roadside. While awaiting clients, the owners played a game of *bao* next to the entrance of their little brick house.

Levocatus was hailed by a boy sitting on a stool on the opposite side of the road. Standing behind him was the barber, who was busy cropping his head with shears. I stood waiting while Levocatus walked over to them. Three bee-eaters landed on a power line next to the railway. Egg-yolk yellow, green, black, white? I could scarcely see their colors in the bright light. Then I felt a hand tugging at me.

Along the way, Levocatus encountered many familiar faces. He approached members of his tribe differently from the way he approached me. As soon as a Sukuma came along, he fell into his role of local celebrity. After all, he was *mganga*, the traditional healer of Nyegezi, a person of consequence. He was adept, like no other, in solving conception problems, though he was very unassuming about it. Genes spoke for themselves, he believed. He was particularly attentive about greeting young women, keeping his head raised and back arched. They in turn made deep knee bends, greeting him shyly, eyes closed all this time, keeping their backs and necks perfectly erect because of the brimming baskets on their heads, which were filled with rice, maize meal, beans, and fruit. The head of a giant Nile perch protruded from one of the baskets. A mouth and a cheek supporting a coolly leering eye stuck out from under the colorful piece of cloth in which the fish was wrapped.

How was Levocatus, and how were his three wives and more than thirty legitimate children? One of his wives, Theresa, had had brain malaria, he said. She wasn't well any more and frequently wandered off. Levocatus didn't have much control over her these days. "It's difficult to live with women who live so far apart," he said. Difficult for the women too, I thought. In the past they would have lived together on one plot and probably helped each other. Today, the wives of men like Levocatus lived on different plots as a gesture of goodwill toward the Catholic church. I would have loved to hear what his wives thought about this diaspora in honor of monogamy, but I never managed it. For these peasant women, a man—certainly a white man—was a different species. Was he still cooking for the missionaries, I inquired? Levocatus assented. Only by combining different professions and activities could he support his extended family. He was working as a mission cook, watchman, psychologist, traditional healer, distiller of spirits, and cannabis farmer, as well as trading in everything that generated money.

The derelict carcass of a bus lay along the roadside, lackluster and rusted. Three words, painted in elegant red letters on the back of the bus, could only be deciphered with difficulty: *Shauri a Mungu*, On my honor! Parched stalks of maize waved back and forth in the bus. A snow-white goat appeared in the doorway at the front. It looked left and right, then dismounted like a seasoned passenger, followed by the other members of its company, which included three other goats. They found what they were looking for by the roadside.

Midge Oil

We approached Mkuyuni, a small market town midway between Mwanza and Nyegezi, and decided to rest. The wall of a dilapidated wooden shed where lemonade was sold bore unpracticed handwriting: Miami Bech Hotel. I was out of breath and could scarcely drink. A passionflower plant, whose blue flower and white heart contrasted sharply with the corrugated iron roof, had sprawled up the wall and across the roof of a neighboring market stall. The saleswoman, a matron who barely fitted into the stall, beckoned to me, pointing proudly to her supply of canned margarine. The cans were only half visible because of the line of lace panties hanging in front of them. On the plank below lay jars of Vaseline and Maria tea biscuits.

"Excellent," I said, raising my thumb in approval.

But it wasn't that easy to shake her off. One by one she tried to sell me each of her wares. As soon as she noticed I didn't intend to buy anything, her face clouded over. You don't want any margarine, panties, Vaseline, biscuits? What good are you to me? Sullenly, she disappeared beneath the counter, re-emerging a moment later to hand me midge oil in a small slippery bottle sealed with a cut-to-fit cork. "Take it, wanderer. You'll need it," she said.

"Are there more midges than there used to be?" I inquired.

"There are many, ever so many. We can't sleep at night."

"But are there more than there used to be?"

The woman burst into laughter and asked Levocatus what on earth I meant. I immediately regretted my question, but it was too late.

"Most of the fish that eat midge larvae have disappeared from the lake. Because fewer larvae are eaten, the number of midges might have increased dramatically," I explained. Only now did I realize that she didn't know that a midge had larvae, let alone that their larvae lived in water.

"I seeee," said the saleswoman, writing me off once and for all. She took the midge oil out of my hands and gave the bottle to Levocatus: "Honorable old man, take this oil for your friend. It will do him good."

I was only half listening to Levocatus and the saleswoman. My attention had been caught by two women standing talking in the shade of a covered market. They had come from the butcher. Each had an aluminum platter on her head, bearing the chopped-off head of a cow with a protruding tongue. The women were conversing animatedly, laughing, clutching at each other's shoulders. They circled around each other, making deep knee bends. It was almost like dancing. Only their necks remained as straight as rods. Higher up, the cows' heads wanted to lick each other but they couldn't get close enough.

"*Njoo*, come," said Levocatus. Dried *dagaa*, the little sardines, had been counted and laid out in neat piles near the fish stalls. There were also Nile tilapias with poky heads, and a young catfish with alert little eyes and long whiskers, rattling in protest against its premature death. When it was finally ready to relinquish the struggle, the salesman sprinkled it with water, giving it a reprieve—and the struggle started all over again. On a table next to the tormented catfish lay stacks of Nile perch. Dung flies swarmed above them, bathing their legs in the blood of the fish slices. After searching for some time, I discovered a scorched brown furu, a snail-eater from shallow waters. It had been smoked.

A group of young boys stopped chewing on sugar cane as soon as they spotted me. "A white one, a white one!" they shouted. They took turns approaching me cautiously, touching me and saying laughingly "*TX*," only to run off a moment later, screaming, arms widespread. TX: the letters with which the number plates of expatriates' vehicles began.

We set off again. There was little traffic on the sand-covered path, which was ridden with deep potholes. The landscape had become quieter, more rural. This used to be a macadamized road. Heavy lorries that couldn't drive through Uganda during Amin's reign and for several years after, had been obliged to take this route. The road had become overworked; holes had developed in the tarmac and the sand beneath it had washed away. Even the inimitable drivers of the local Fantastic Express Bus Service were eventually unable to negotiate the slalom course between the meters-deep potholes. Now, lorry traffic was once again passing through Uganda to Ruanda, Burundi, and Zaire, and we were walking along a sandy path. Such a path was much easier to maintain than a highway, but its days were numbered. The moment a president or a pope decided to take this route, a tarmac runner would be rolled out.

"I understand they're making a good living from the *sangara*, the Nile perch. But I read that it is especially a small wealthy group that is benefiting," I said.

"Eh," replied Levocatus, assenting. "They have boats and nets."

"But of what significance is the Nile perch to people like you?"

"Sangara was like an answer to our prayers. What would we have eaten if it weren't for the Nile perch? Sangara saved us, *mkombozi*," said Levocatus.

"Mkombozi?"

Levocatus didn't know the meaning of the word in English. I spoke to several passers-by and asked them. "*Mkombozi*," replied a pot-bellied man with a fleshy head, "*mkombozi* means 'savior' ... eh." He beamed with delight, adding needlessly: "So that now you know the meaning of this word."

"Are you from Japan? Ah, from Europe. I'm so sorry. Unfortunately, you look like a Japanese." He burst out laughing and turned the palm of his right hand upward.

I clapped my right hand down onto his and thanked him elaborately.

"*Karibu Tanzania*, welcome to Tanzania," he concluded.

Minutes later I heard Levocatus still softly mumbling: "Savior, savior."

Beautiful Shoes

Boulders were scattered whimsically across the landscape like loaves of bread. Swallows darted swiftly between them in search of prey. Meters-high euphorbias stood like great candlesticks near the boulders. Should I go to the rock paintings? They used to be near here. Once, by chance, when searching for leopards and caracals, I had come across symbolic paintings on an overhanging rock. They were not animal or human figures, but abstract symbols. I had no idea what they had originally signified or how old they were, but I was inclined to make an offering before proceeding to the lake: breadfruit, a mango, and limes, surrounded by a wreath of flowers, and next to this, on a banana leaf, a slice of *mkombozi*. Then I muttered a prayer asking the savior to make a quick exit before the ecosystem burst and the lake was transformed into a lifeless desert. "Hey, dreamer," said Levocatus, as he took my hand and pulled me along.

For most of the journey, hills hid Mwanza Gulf from view, but I smelled the lake and once in a while caught a glimpse of it. When I heard the call of the fish eagle, I could scarcely contain myself any longer. I wanted to go to the water.

A small pickup truck filled with people passed us at the turnoff to Nyegezi. They laughed, waved, and the truck stopped. Levocatus and I couldn't see anything through the whirling sand. We squeezed our eyes shut, shielding them with our arms. A moment later, laughing faces loomed up out of the cloud of dust. It was colleagues from the institute: Kabika, Bahati, Kitabwa, Jane, Kurwa, and Maembe. Hearty greetings followed and I was bombarded with questions. How is your wife? Did you bring the clutch? How many children do you have?

The driver was more intrigued by my shoes than by my limited reproductive success and commented: "*Viatu vizuri*, beautiful shoes, exceptionally beautiful shoes."

"'Through the Bush' they're called," I replied.

"Very good," said the driver. "I must go through the bush. Maybe we can cooperate."

Levocatus and I declined the offer of a ride and continued on foot. Overwhelmed by the number of familiar faces that had emerged from this one vehicle, I worried silently whether I had enough *zawadi,* or gifts, with me to give everyone the feeling that I had not forgotten them.

Luguru Hill could now be seen in the distance but I couldn't distinguish the buildings of the mission retreat. Only a vague stripe—the pillar on which the church bell had once hung—was visible from here.

It was heavy going on the loose sand. I was surprised by Levocatus's youthful gait. He must have been about sixty, and he was still agile, tawny-colored, and muscular. I trudged along slowly behind him. "*Labda utapata mtoto sasa*, maybe you will eventually have a child," he said. I was amazed by the fanaticism with which the Tanzanians worked on their reproductive success. I had forgotten about it.

A pied kingfisher flew low overhead clutching a dagaa in its beak. We sat down, looking out on to a steep quarry, which looked as if it had been fired at with a cannon. The wall was riddled with bullet holes. A second glance revealed that the distribution of holes across the wall was not at all haphazard but regular. There was system in their placement. The holes were the entrances to the kingfishers' nests. Dozens of kingfishers, mostly males, were chattering in a tree next to the quarry. They could be recognized by the small band of black feathers encircling their white necks like a necklace.

Almost as many males as females emerge from the eggs, but by adulthood, males far outnumber females. Females spend more time brooding, and perish in doing so more frequently than foraging males. Black mambas, cobras, monitor lizards, and zebra mongooses form an ongoing threat to the broods and the females. This is why the males come to outnumber the females. Several males with little fish in their mouths disappear in quick succession into one of the holes. A brooding pair is helped in foraging by a third bird. These helpers may be sons or daughters of the pair, or brothers or sisters without mates. Sons and daughters are always accepted as helpers, unlike brothers and sisters. The more difficult it is for parents to supply the young with food, the more inclined they are to accept help from third parties. It is a flexible helper's system. Helpers related to the brooding pair make a virtue of necessity. By investing in cousins, they are making an effort to disseminate their genes. Even though this is less efficient than if they brooded themselves—because the degree of kinship between cousins is half as great as that with one's own young—for a young or single bird it is better than nothing and it undoubtedly benefits from it in some way.[150] When the next kingfisher returned from the lake, again with a dagaa in its beak, I decided to stay and watch. "*Sawa, sawa*, that's all right," said Levocatus, as he stood up and brushed the red earth from his trousers. "See you in a while, on the hill." The next four kingfishers also returned with dagaa in their beaks, but there wasn't a furu to be seen. Before the arrival of the Nile perch, this would seldom or never have been the case. Even in the foraging behavior of these birds you could see that the ecosystem had changed.

The Institute

Dozens of girls were working in the school field with a *jembe,* or spade. They were wearing green pleated skirts and immaculate white blouses. "*Habari za kazi*, how's it going?" I asked a girl who was staring at me as she leaned on her *jembe*. She was large and had a sturdy build: massive loins and heavy thighs. Then I was struck by a nasty memory. This was exactly the place where the Land Rover had stood. It had been late afternoon. As was the case today, the girls had been obliged to work on the school plot. Three older men, government officials, had sat in the vehicle with a woman, the director of the school. The men chose a girl for the night from the stooping crowd. If the chosen girl refused, she was sent from school. If she became pregnant, her lot was no different. The chance of becoming infected with AIDS played no role in the story. It was 1984 and nobody in the region knew that an epidemic was spreading around them like wildfire.

I strolled toward the institute. Machota, one of the guards, opened the iron gate. He looked at me as if he'd seen a ghost, uttering my name several times in a muffled voice. He wasn't a Sukuma but a Kuria from northern Tanzania, recognizable by the ear lobes that were rent open like key rings. A second *askari* emerged from the guard's hut. He stretched, then picked up his truncheon.

The institute made a desolate impression. It was still early but most of the employees had already departed, after leaving behind conclusive evidence of their presence in the form of a signature. They would now be trying to extricate overdue salaries from the necessary authorities in town, a lengthy process requiring great perseverance.

As I crossed the inner courtyard, I interrupted a gathering of the cold-blooded. Dozens of lizards and agamas shot away under the fish nets that were lying on the ground drying. It quickly became evident what all these reptiles lived on. Under one of the night lamps lay a mountain of tiny winged corpses of nonbiting midges. There were more than ever before. The lake was leaking, was my first thought. Energy was spouting out of the system. In the old ecosystem most of these midges would have met their end in a fish. The mountain of dead insects confirmed my expectation that the midge population would increase now that the predators of their larvae had more or less disappeared.

A haze hung above the lake, columns with vaguely defined contours, hundreds of meters high. It was swarms of midges that could be blown away just like that. Never before had I seen such enormous clouds of them, except near Lake Malawi, where they were harvested. There the people rolled them into balls and ate them.

Most of the midges belonged to the genus *Chaoborus*. Their larvae lived in the muddy lake bottom during the day, migrating after sundown into the water column to eat zooplankton. The chance they would be caught by sight-dependent predators was relatively small after sunset. The life cycle of the *Chaoborus* species, which lasts about two months in the tropics, is geared to the amount of moonlight. Around new moon, when the quantity of night light is minimal, the larvae move into a pupal stage, leaving the water as midges. Mating takes place, the females lay eggs that end up in the water, and the cycle begins all over again. In a disturbed system the midges are no longer an intermediate link. Instead, they form the final link in the chain. Several food chains in the changed ecosystem gradually occurred to me, but it wasn't yet clear to me how they were linked.

Mhoja, Kabika, and Elimo waved and called from the *Sangara*, which lay several meters offshore. I was lucky. They were going fishing today and I was welcome to join them. The crew prepared the *Sangara* for departure. They picked me up with a canoe and once again there was an extensive greeting ritual.

"*Mbwana Tesi*. Did you receive my letter?" asked Mhoja.

"I certainly did," I said, assuming he meant the shopping list.

"Your fish have left," said Elimo, laughingly.

"Yes, I know," I replied, struggling to keep my balance as I stepped into the canoe.

It had happened so quickly. It was only in 1985 that I had first realized the Nile perch was wreaking havoc in Mwanza Gulf. The species of furu that I had been catching for years, week in and week out, in the northern part of Mwanza Gulf had disappeared very rapidly, to be replaced by several species from deeper waters. I had left for the Netherlands in 1986 but the work had been continued by other researchers. By 1988 all the furu in the open waters of Mwanza Gulf had disappeared, with the exception of a few zooplankton-eaters. How would the system be now, one year later?[151]

"Where would you like to fish? You can choose, because you didn't forget us," said Mhoja.

I asked the crew to make a sweep across Station G. The anchor was raised. The kingfishers sitting on the edge of the wheelhouse flew—chattering—back to shore and the boat chugged out of Nyegezi Bay, heading for my old haunt.

The wind began to blow harder. The clouds of midges swarming above the lake disappeared. Parallel white stripes traversed the rippled water's surface in the same direction as the wind. A group of dirty white pelicans landed behind the trawler, bobbing around in its wake, patiently awaiting fish.

After we'd made the sweep across Station G, the cables of the net were winched up and a short time later, the net emerged from the water. Dripping, it swung back and forth on the gallows. Before the Nile perch era, the filled death trap had been shaped like a giant flower bulb. Normally, it had contained hundreds of thousands of furu, a few dagaa, and several lungfish, catfish, and Nile tilapias, as well as the occasional elephant snoutfish, spiny eel, or other less common species. This time the death trap was bulging unevenly at the front. The fish were large but distinctly fewer in number. Mhoja gave a tug on the rope that sealed the death trap and jumped aside. While the fish cascaded onto the deck, he did a little jig. "*Mbuzi, mbuzi*, a goat, a goat," he shouted, radiant. The catch included a Nile perch weighing more than forty kilograms. It gave me so much pleasure to see Mhoja acting in this way that I forgot my revulsion at the sight of all these dead fish.

The composition of the catch was totally unfamiliar. If I had been shown a photograph of it in the Netherlands, I wouldn't have believed it had come from Lake Victoria. Nile perch, large and small, lay on a thick, soft bed of little brown prawns. Silvery dagaa shone here and there, and there were a few Nile tilapias. And that was it. No lungfish, no catfish, no elephant snoutfish, and not a single furu. A frightening deterioration of the ecosystem. Within a single decade, the differentiated biotic community that had coevolved over a period of at least fourteen thousand years, and perhaps even hundreds of thousands of years, had changed into an impoverished mess. But something wasn't quite right. How was it possible that so many Nile perch were still being caught when their primary food source, the furu, had already disappeared years earlier?

A Prawn Develops

An article, cosigned by our group, was published in 1985 in which it was predicted that the introduction of the Nile perch would have disastrous consequences.[152] The introduction of this predator was considered a grave error from the ethical, scientific, and even technological point of view. In the past there had been heavy fishing of furu, most of which consumed phytoplankton and organic waste. From sunlight via phytoplankton and organic waste to fish. A shorter and more efficient food chain didn't exist. How could people be so arrogant as to meddle with a system that, in the course of such a long period of coevolution, had moved toward a more or less stable equilibrium? That had been the thrust of the article.

During the mid-1980s, more than 80 percent of the fish biomass of Nyanza Gulf in Kenya consisted of Nile perch: "An atypical and clearly unsustainable role for a predator at the top of the food chain." Why was it alarming that a top predator formed the largest part of the biomass? It was because the food pyramid had been turned upside down. In the odd instance, where prey, in relation to their predators, have an extremely short generation time, it needn't be disastrous, but this wasn't the case here at all. Every ecologist would panic if he saw vast herds of lions in the Serengeti running after the last existing antelope. This was exactly what was threatening to occur in Lake Victoria. The Nile perch population would collapse. The destruction of the indigenous fauna would have been pointless, even from the economic and technological point of view. It was the essence of stupidity.

This is the way we viewed the situation for many years, but in fact, it turned out quite differently. Every time we predicted something, the ecosystem surprised us by taking an unexpected turn. One such surprise was when a previously unimportant energy flow suddenly became of vital importance. In this case it was the development of the prawn.

Organisms in an ecosystem can be divided as follows.[153] First there are the plants, the primary producers. Strictly speaking, they are not producers of energy; they only transform it, during the process of photosynthesis. With the help of solar energy, carbon dioxide and water are converted into energy-rich carbon compounds and stored in the plant in this form. Oxygen is released in the process. The energy-rich carbon compounds, such as sugars and starches, subsequently become available to the primary consumers. These second-level organisms graze on the primary producers. Organisms of the third level—the secondary consumers—graze on the primary consumers and so forth, up to the highest trophic level, that of the top predators. It is striking that the number of links in a food chain is seldom more—and usually less—than five, irrespective of the wealth of species in the system. The number of the trophic level indicates the number of links in the food chain separating the level at hand from solar energy.[153]

In 1942, the ecologist Lindeman demonstrated how to calculate the total energy content of organisms at a certain trophic level. He introduced the concept of "ecological efficiency." This is a measurement for expressing the relationship between the production of chemical potential energy at a certain trophic level and the loss of energy that inevitably takes place. After all, an organism must maintain its structure, and grow and reproduce. Moreover, predators are responsible for a loss of energy at the trophic level of their prey.[153]

Lindeman's approach facilitated an objective comparison of all ecosystems: in water or on land, young or old, tropical or nontropical. For ecosystems in a state of equilibrium, approximately 90 percent of the energy disappears at each step in the trophic ladder. If energy production at the first level is assumed to be 100 percent, then 10 percent of this energy remains at the second level and only 1 percent at the third. The quickly shrinking energy flow of an ecosystem can be rendered graphically in the form of a pyramid, in which the broad base consists of the primary producers and the top consists of the predators of the highest trophic level. These top predators, which as a rule are large and scarce, are not preyed upon by other organisms.

In the same way as the energy flow through an ecosystem can be established, it is possible to calculate the total biomass embodied by successive trophic levels. Normally, primary producers represent by far the largest biomass, followed by the biomass of detritus-eaters, which extract the last remnant of energy from dead plant and animal tissue. The top predators represent the smallest biomass. The reverse pyramid of the energy flow and biomass explains the concern about the dominance of the Nile perch in the fish community of Lake Victoria. But the prediction that the golden age of the Nile perch could not outlast the supply of indigenous fish turned out to be untenable.

Nile perch in the open waters continued eating furu until there were virtually none left.[154] But where then were the thousands of starving

Nile perch floating moribundly on the water's surface? Why hadn't their numbers declined rapidly? Why had the predicted collapse of the Nile perch population not taken place?

It was because there had been an explosive increase in the population of a prawn, *Caridina nilotica*. The crew scooped up one container after another from the nets, all filled with prawns. Where did these unlikely quantities come from? In the old ecosystem, I had occasionally stood eye to eye with a prawn but it was usually only one. If someone caught more than three prawns in a day, it was newsworthy. Most of the prawns I came across were in the stomachs of prawn-eating furu, which were specialists in tracking them down. And now, millions of prawns lay spread across our deck.

I cut open a dead Nile perch and examined its stomach contents: prawns. Its last meal had consisted of prawns. This explained why the Nile perch population hadn't collapsed. Since the disappearance of the furu, prawns had become its main prey.

The mass appearance of prawns in the open waters was indirectly attributable to the introduction of the Nile perch. The large group of furu that lived primarily on organic waste at the bottom of the lake but wouldn't turn their nose up at a prawn had been eradicated by the Nile perch. Moreover, the species of furu that were specialized in eating mature prawns had been obliterated. Prawn predation had diminished rapidly and a Nile perch population explosion had followed.[155]

Nile perch eat dagaa as well as prawns, and cannibalism is also rife.[156] They frequently assault younger members of their own species. Even young Nile perch six to ten centimeters long are cannibalistic and consume two- or three-centimeter-long Nile perch on a large scale. The Tanzanians quickly learned of this and many were horrified. They didn't want to eat Nile perch for fear that cannibalism was contagious.

That same evening I learned that the dagaa, along with the prawns, had also increased greatly in number. Mwanza Gulf—a sea of tiny lights—resembled a city by night. The dagaa couldn't resist a concentrated source of light and the fishermen exploited this around new moon by luring them with lamps and subsequently catching them.

The increase in the number of sardines was also a result of the introduction of the Nile perch. The zooplankton-eating furu had been virtually

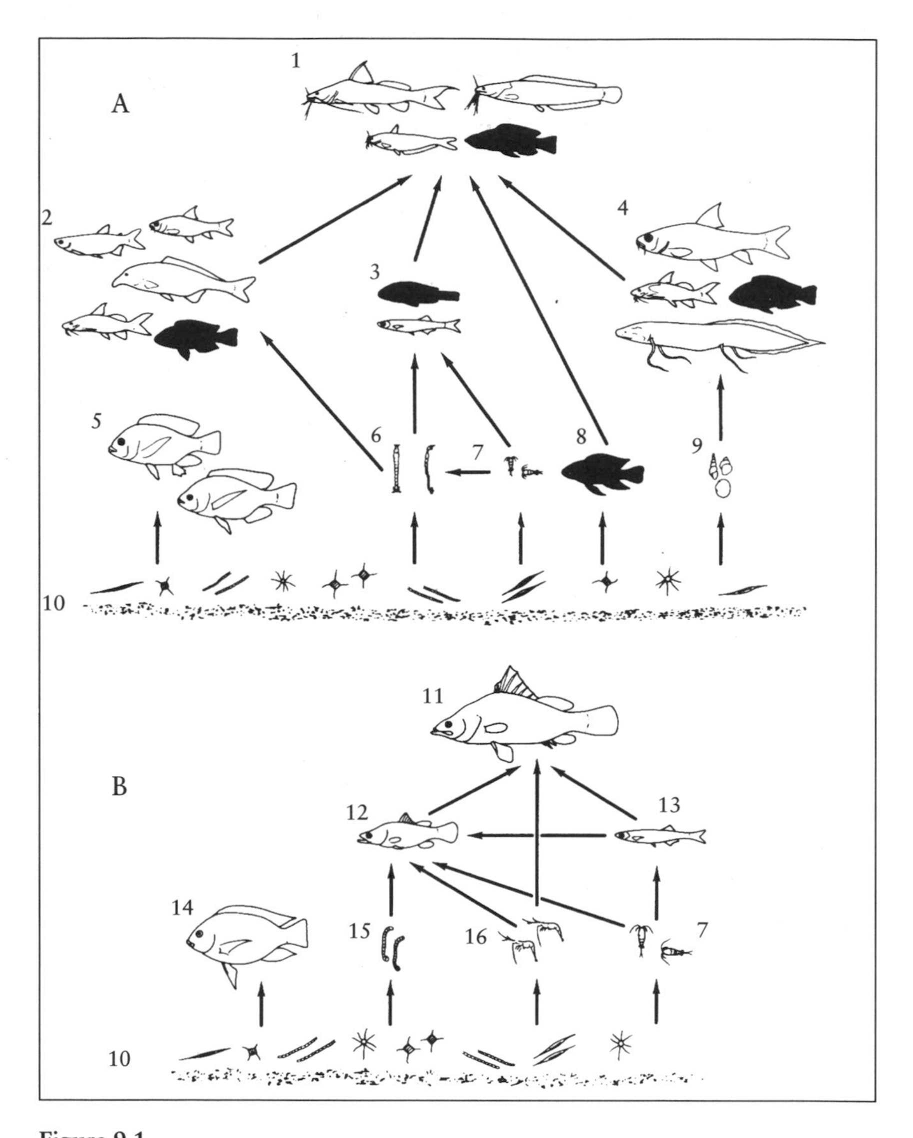

Figure 9.1
Simplified representations of the food web in Mwanza Gulf before and after the proliferation of the Nile perch (A. during the 1970s; B. in 1989).

Only those organisms that occur in the diet of the most common species of fish are depicted. Furu are shown in black. The numbers correspond to the following groups: 1. fish-eaters (catfish and furu). 2. insect-eaters. 3. zooplankton-eaters (dagaa, the little sardine, and furu). 4. snail-eaters (including the lungfish and furu). 5. algae-eaters (indigenous tilapias). 6. mosquito larvae. 7. zooplankton. 8. mud-biters (organic waste- and phytoplankton-eating furu). 9. snails and shells. 10. organic waste and phytoplankton. 11. Nile perch. 12. young Nile perch. 13. dagaa, the little sardine. 14. Nile tilapia. 15. mosquito larvae. 16. prawns.

wiped out and the sardine population had increased.[157] But I don't know quite how. The quantity of zooplankton might have increased after the zooplankton-eating furu had disappeared. But this wasn't necessarily the explanation. Sardines have a different reproductive strategy than that of the furu. They offer no parental care to the brood and fry, unlike the furu. They produce a relatively large number of eggs per clutch, and, following fertilization, give them no further attention. In an unpredictable environment such as the disturbed ecosystem of Lake Victoria, this might have represented a more favorable strategy than the energy- and time-consuming hatching of a small number of young that would most likely perish before they themselves were able to breed.

In addition to Nile perch, sardines, and prawns, our catch also yielded a few Nile tilapias. Like the Nile perch, Nile tilapias had been introduced during the 1950s, eventually supplanting the indigenous tilapia species.[147]

Nile perch, sardines, prawns, and Nile tilapias. These were the main pillars of the impoverished system. It became increasingly clear to me why the original food web had collapsed. These were the most important role changes that had taken place:[158]

1. The Nile perch had come to occupy the position in the ecosystem of the fish-eating furu and the furu-eating catfish (*Bagrus dokmac* and *Clarias gariepinus).* Noteworthy here is the fact that a single predator had taken the place of more than one hundred species of predators.
2. The prawn (*Caridina nilotica)* had replaced the prolific detritus-eating furu (at least thirteen species).
3. The dagaa, the little sardine (*Rastrineobola argentea),* had taken the place of the more than twenty species of zooplankton-eating furu.
4. The Nile tilapia (*Oreochromis niloticus)* had replaced the indigenous tilapias (*Oreochromis esculentus* and *variabilis).*

An important change in the ecosystem has not yet been mentioned: the disappearance of the algae-grazing fish.

The role of the detritus-eating furu had largely, but not entirely, been usurped by prawns. The detritus-eating furu left the bottom of the lake at night to graze on the phytoplankton. But prawns don't eat living phytoplankton.

Moreover, there was an important group of pelagic algae-eaters: three species, which together formed 18 percent of the biomass of the furu. The phytoplankton-eaters had also disappeared, like the detritus-eaters. As a result, the ecosystem had lost its most important group of primary consumers. Vertebrate algae-grazers were absent in the new ecosystem, although planktonic crustaceans that ate phytoplankton were present.

The role of the insect-eating furu had not been—or had scarcely been—taken over by other organisms. This might explain the increase in the number of midges. The same probably also applied to snails and bivalved mollusks. The mollusk-eaters had been wiped out in the open waters, and during the past few years, enormous quantities of mollusks had been caught.

The effects of the introduction of the Nile perch cascaded like a waterfall through the food web and were visible right up into the highest regions of the food pyramid.[155] As already mentioned, the diet of the kingfishers (*Ceryle rudis)* had changed, but the same also applied to the diets of various species of cormorants and the otter (*Lutra maculicollis)*. Otters used to eat mainly furu; now they had switched to Nile tilapias.

Beginning with the algae and detritus at the bottom of the lake, the most important food chains had been structured as follows:[158]

- via algae and detritus-eating furu to fish-eating catfish and fish-eating furu;
- via zooplankton and insect larvae to furu and sardines;
- via insect larvae to various species of fish;
- via mollusks to various species of fish;
- directly via algae and detritus to species of tilapia.

In the new food web the most important chains worked as follows:

- via prawns to the Nile perch;
- via prawns to the juvenile Nile perch and thus, via cannibalism, to the larger Nile perch;
- via insect larvae to the juvenile Nile perch and thus to the larger Nile perch;
- via zooplankton to sardines and the juvenile Nile perch and thus to the larger Nile perch;
- directly via algae and detritus to the introduced tilapia (*Oreochromis niloticus)*.

Compared with the original system, a major shift had taken place: a fish community consisting mainly of primary consumers (algae and detritus-eating furu) had been transformed into a community in which the Nile perch, a top predator, dominated. The Nile perch operated primarily as a secondary and tertiary consumer.

It has long been known that the introduction of alien species—and overfishing—can have a major impact on the food chains of ecosystems.[159] For example, during the past century the ecosystems of the North American Great Lakes, such as Lake Michigan, have changed in a way that closely resembles the recent changes in Lake Victoria. The ecosystem of Lake Michigan was totally disrupted by overfishing and by the introduction of nonindigenous fish. The greatest damage was caused by the sea lamprey (*Petromyzon marinus*). Endemic lamprey species were found in the lake, but they stayed small even into adulthood. The sea lamprey grew much larger and had a devastating effect on the lake char (*Salvelinus namac*) population, already under pressure from commercial exploitation.

As in Lake Victoria, the introduction of exotic fish-eaters had far-reaching consequences for the structure of the food web. The fisheries collapsed. It became necessary to manipulate the ecosystem in order to curb the influence of the sea lamprey and to successfully launch new types of fisheries. By artificially limiting the number of sea lamprey, the indigenous fish-eaters were given an opportunity to recover.

Today, the food web of Lake Michigan is totally dominated by introduced species. The ecosystem has to be continually manipulated to maintain a productive equilibrium. The countries around Lake Victoria lack the necessary infrastructure to allow such an active manipulation of the ecosystem. Removing the Nile perch from the lake is already totally out of the question. The fisheries of Lake Victoria have not yet collapsed, but the threat is imminent. The ecosystem is far from stable. Ecosystems that have been greatly simplified by humans are often only productive for a short period. If this were also to happen in Lake Victoria, it would be the umpteenth example of shortsighted policies. Ethical or aesthetic arguments for the preservation of fauna have no chance at all in the face of short-term economic perspectives. Certainly not in poor African countries, in which populations barely manage to survive. Conservationists are being compelled to express the value of an area in terms of money and if they refuse or are unable to do so, such an area is doomed.

10
The Oracle

Jinja, Uganda, 1992

The journey again: hanging around, waiting for connecting flights. In a bleak waiting room at Nairobi airport, a small group of Africans was having a little party. A ghetto blaster on the gray linoleum was playing at full volume. M'bilia Bel's voice, shrill and coarse. Two young men and a woman with a child on her arm were dancing animatedly. An older man was sitting on the ground playing a three-stringed instrument. None of his music was audible because of the ghetto blaster, but he didn't seem to mind. The sky turned a glaring yellow—burning stripes, in the colors of the baggy shirt of one of the dancers. Now he was on fire, now he wasn't.

It would be at least another hour before the flight left for Entebbe. I removed a few articles on Lake Victoria from my bag and started reading them. The quantity of blue-green algae had increased greatly during the past few decades.[160] The lake had become more nutritious for these organisms because of the influx of nutrients as a result of deforestation and possibly, acid rain. The lake's surface was covered with endless fields of blue-green algae. Dying algae sank to the bottom and were decomposed by oxygen-consuming bacteria. Fifty to 70 percent of the lake was currently deoxygenated all year round.[161] In the past, during long periods with little wind, an oxygen-deficient layer had developed in deep water but the separation had disappeared as soon as the wind mixed the water again.[162] A subject of discussion was whether the increase in blue-green algae was also a consequence of the disappearance of algae-grazing fish from the ecosystem, an indirect effect of the introduction of the Nile perch, which had consumed all the algae-grazers. Because it was also conceivable that the eutrophication of the lake might have occurred in exactly the same way without the Nile perch. Now to make sure I didn't miss my flight. Most of the passengers were already on their way to the aircraft.

From Entebbe by bus to Jinja. A long ride over dusty roads. Savanna, little houses with corrugated iron roofs, cows on the road. The silhouette of an eagle in the bright light. Disembarked at Nile View Hotel, a building with pink-stuccoed walls dating from colonial times. It was perched on top of a hill. I was greeted by several Ugandan biologists. I left my baggage and walked into the luxuriantly overgrown garden. Dozens of shades of green, and as many shapes of leaves. Sunbirds with heads as red as if they had been dipped into a jar of paint. Dry rhythmic ticking and a restrained rattle from the hedge. Disjointed clucking from high up in a tree and loud hissing from a shrub, followed by a welcome duet. The sounds of birds, unmistakably tropical.

Hands raised, I leaned against the wire of the meters-high fence encompassing the garden. Directly below flowed the White Nile, whose source—Lake Victoria—was not far from here. Was it in 1857 or 1858 that Speke had seen the lake for the first time? His fellow-explorer, Burton, had stayed behind in Kazeh, present-day Tabora, a center of the slave trade, traces of which were still visible in the form of the old mango trees found along the road leading there. These would have sprouted from the mango stones left by the slaves as they passed on their journey. Burton had been skeptical when Speke told him he had seen the source of the Nile. An enormous gush of water, that must be it. Not enough evidence, thought Burton, because Speke had not found the place where the Nile flowed from the lake. But this was not surprising as he was on the wrong side of the lake, near Mwanza.

Had seventeen years already passed since the antitravelers Barel and Anker had sat moaning a few kilometers from here, endlessly reiterating that there was no reason for them to be in Africa except for the threatened furu with their irresistible heads? More than ten years ago I too had left for Mwanza, in the hope that it would be illuminating to spend some time in an undisturbed tropical ecosystem. During that time the system had changed in a rapid tempo and new laws had come into effect in and around the lake.

On the other side of the broad river lay rolling hills without buildings. I pushed my nose through the wire, as far as I could into Africa. Behind me, the hum of conference participants: conservationists, ecologists, limnologists, economists. Coming together to talk about the future of the lake.

How to convince politicians of the importance of nature conservation when it generated no money? Commercial fishing was booming. A great deal of money was being earned, some in foreign currency, of which the countries surrounding the lake had an acute shortage. As long as this was the case, the introduction of the Nile perch had at least had the effect technologists were hoping for, even though the indigenous fauna had paid a price for it. But would profits continue to be so high?

What of the original fauna could still be saved?[163] A great deal, no doubt. Indigenous tilapias, elephant snoutfish, catfish, and other species were biding their time in hatcheries around the lake, waiting until they could return. But what had to happen first? Did reserves need to be built to protect the furu against the Nile perch and commercial fishing? Which was more vulnerable: one large reserve or several smaller ones?[164] Was there any point to breeding furu in captivity in the West? How many species and for what purpose? To preserve them like living fossils for years to come in zoos? How real was the possibility of their ever being let out again in this lake or in the other small lakes in the area? Which species were to be allowed into the ark and which would we condemn to extinction? What a profession![165]

There was no time to get used to Africa. Long before I realized where I was, the meeting would be over and I would be airborne again. It was strange that not even half a day had been allocated for visiting the lake. Inlets had become overgrown with water lilies and water hyacinths, and populated with egrets, jaçanas, monitor lizards, and pythons. I had only heard of these apparent paradises, but there was no time to see them. Our job was to talk and write reports, to clean up the mess, quickly and cheaply.

My legs were shaky after a thirty-six-hour journey, and I hesitated about plunging into the crowd. The social hour, an obligatory part of every scientific meeting, was in full swing. If only I could sleep for a few hours first. I felt a tap on my shoulder.

"What you said about those algae-eaters. That article ... I could not get much out of it ... Oh, I'm sorry." The man who had addressed me offered his hand. “Do you remember me? ... The Bujumbura meeting. Hi ...”

I bent over to tie my shoelace and looked at the name tag pinned to the jacket of my amply chested colleague: Nizar Nyaskebo.

“Nizar! Now I recognize you!”

The circus had started. Weren’t those the “refugees” on the other side of the garden who had stayed at my house? Was Melle here already? I hadn’t seen him for years.

“Even if the Nile perch had swum with a respectful detour around every algae-eater they came across,” Nizar rattled on, “the algae would still have proliferated.”

Eutrophication of the lake had started before the beginning of the Nile perch era, owing to the influx of nitrates and phosphates, and to acid rain and perhaps changes in the climate. That was why fields of blue-green algae were growing on the water’s surface.[166]

A young woman without a name tag squeezed her way through the crowd. She was wearing a shiny, dark yellow, ankle-length dress, and moved gracefully in the direction of the barbecue, where a cook was working. She saw me looking at her and approached me a short time later. “I stay in room 17,” she whispered, tapping me on my shoulder. “I’ll do my best. See you later.” Hips swaying, she disappeared again, swallowed up in a sea of blue jackets.

Nizar leaned toward me and said: “This lady is prepared to do anything that is allowed between man and woman.” He giggled and pressed his index finger against my chest. “A man needs a woman, it’s as simple as that. Don’t look so uncomfortable. *Unaogopa*, or are you afraid? Sometimes I just think sex is more integrated in our society than in Britain.” Nizar turned around and said: “There goes another beautiful specimen, but perhaps you’re more interested in saving the last goldfish?”

Kishamawe and Maembe strolled toward us. Water hyacinths and oxygen. That’s what they were working on. I had met them at a previous meeting. They had a glass in one hand and a bottle of beer in the other.

“The lake is a memory,” said Maembe.

“Kishamawe,” asked Nizar, “when did the first water hyacinths appear on Lake Victoria?”

“In 1990, via the Kagera River. They spread out very quickly along the shoreline. They’re suffocating the inlets.”[167]

I was only half listening and looked to see if I could spot Melle among a group of participants strolling into the garden.

"Within a century, the lake will be a stinking swamp, as it once was."

"It might turn out quite differently," I suggested.

"Oh, yes?" said Maembe. "Have you ever smelt the lake when masses of fish have died? When thousands of rotting Nile perch float to the surface? This has been seen many times during the past few years.[168] They have suffocated, been overwhelmed by a giant bubble of deoxygenated water pushed up from the depths to the surface."

"Perhaps that also happened occasionally in the system when it was still undisturbed?"

Maembe didn't hear me: "If such a bubble of water is blown into the littoral zone one day, the last few hundred species of furu will also die out. Just like that."

I felt dizzy. The bitter taste of quinine pills. I excused myself, pursed my lips together, and retreated to the WC. Ten years earlier I had arrived in Africa vomiting. Now, scarcely having landed, I was vomiting again. Exhausted, I returned to my room and fell onto my bed. "Anything that is allowed between man and woman." That one sentence kept me going.

When I awoke the next morning, the day was already into its fourth hour. The dining room was empty—breakfast had long since finished. Fortunately, the minister had already delivered his opening speech. One of the organizers was now giving his talk. A table was projected onto the screen: hundreds of not-quite-legible figures. Melle was sitting at the back of the dimly lit room. I walked toward him and put my hand on his shoulder.

"Just arrived?" I whispered.

"Last night."

I pulled my chair up next to his.

Melle nodded in the direction of the speaker behind the wooden podium and said: "Just before you came in, he said that the extinction of the furu species flock is the first mass extinction man has consciously seen, felt, and recorded."

Melle wrote the words "seen," "felt," and "recorded" on a piece of paper, encircled them, and slid the paper toward me.[149]

I hardly registered a thing. I had taken it upon myself to draw an illegible number as clearly as possible. Every now and then I made a note to remind myself of something from a certain talk: one planet, one experiment (with thanks to E. O. Wilson) /ni...tro...gen.../ ... the oracle had spoken.

The oracle? At twelve o'clock sharp the chairman announced the unintelligible name of the next speaker. A small thin man with gray hair walked to the front. He tried, unsuccessfully, to show a slide. The projector clicked—the slide stuck. Too thick. He bowed apologetically to the audience and disappeared behind the lectern. Rattling sounds were heard.

"Is he from the Japanese group?"

"I don't know. I don't see him here," said Melle, pointing to the list of participants.

The chairman put a cassette recorder on the podium. The little man was still invisible when a hand with an extended index finger appeared above the recorder. There was dead silence in the room. The finger descended. Click. Dzzzt. The hum of the recorder. The noise continued for a while until someone started speaking. A low female voice. Commencing slowly, becoming increasingly passionate in a language unknown to me. The voice stopped occasionally, breathing heavily into the sea of humming, then continued again, impassioned. This was no longer reading but invoking. The motor moaned and groaned as if it might give way at any moment. Might the voltage be too low? Was the Japanese lady presenting her text in English? The chairman walked to the machine and turned the volume up. It didn't help. He turned the volume down again. After eleven minutes he placed a white flower on the recorder. The female voice continued ecstatically, as inaudibly as ever. Three minutes later the chairman placed another flower on the recorder. Now the voice faded away, no longer drowning out the humming. The hand reappeared above the recorder. The index finger straightened and descended. Click. Dzzzt. The man emerged from behind the podium and bowed deeply to the audience. Then, punctuated with long pauses, he uttered three words: "No ... questions ... please," after which he bowed and disappeared into the falling darkness.

The next evening I strolled through the corridors of the hotel, feeling giddy from the dozens of speeches. I wanted to leave. But Jinja was supposed to be unsafe at night and what is more, there was nothing to do there. The conference participants were sitting at the bar, forever talking, as if their survival depended on it. They were talking at the dining room tables. And there were groups of people standing in the corridors—also talking. I didn't want to hear or say another word. I was ready for room 17, only curious about one thing now. I drank a cup of coffee at the bar and walked up the stairs, along the outer corridor to room 17. I couldn't find it. I saw 17A, 17B, and 17C, but no room 17. I turned around at the end of the corridor and retraced my steps, wandering into the garden. The top part of the room doors was visible from there. In my left trouser pocket, I counted three vacuum-packed sheaths, my guardian angels. It was time. I returned and knocked on room 17A. The door was unlocked and someone opened it.

"Oh. No. Sorry. Wrong room," I said.

The little man of the oracular female voice was standing in the doorway. He chuckled and bowed. He wanted me to come in. The room was dark. A slide was projected onto the white wall. The man pointed to the slide and said: "Furu." He closed the door.

I sat down reluctantly on the floor with my back against the edge of the bed, and tried to concentrate. Unmistakably furu, he was right. Furu from Lake Victoria. But I didn't know this one. Strange. Seen them a thousand times and still a new one. A strong feeling of familiarity and at the same time, surprise. Was I dreaming?

"Never seen that one," I said, shrugging my shoulders.

The man chuckled and showed the next slide. Again, that strange feeling of familiarity in the absence of a name.

With each new slide the man nodded in the direction of the image and as soon as I said I had never seen that species before, he chuckled again. Perhaps I had known these fish but forgotten them? I was confused.

Next slide: "But that's ... from Mwanza Gulf!" I jumped up and pointed, unnecessarily, to the image of light on the wall.

The little man picked up a piece of paper and wrote: "I catch these furu in Mwanza Gulf."

"Is that what you were trying to tell us yesterday morning?" I asked.

The man assented, nodding furiously. He lisped, clucked, and gurgled. After being quiet for a moment, as if mustering all his resources, he said: "Furu come back."

Postscript

The furu species that have never been seen before in Mwanza Gulf might have entered the gulf area from the open water, but this is not likely. In that case we would have come across them on one of our many journeys outside Mwanza Gulf. Nor are they river-dwellers. The most probable explanation at the present time is that different species are hybridizing. A male and female from different furu species can produce fertile young because of strong genetic similarities between species.[59] In principle, new species can also emerge in this way, as long as hybrid descendants do not rebreed with parental species.[59] The solutions developed by the unknown forms to survive the onslaught of the Nile perch are still a mystery. Nor can the future of the ecosystem be predicted. But it is not inconceivable that a new species flock will eventually emerge. A new fanning out of forms that are resistant to the onslaught of the Nile perch.

Epilogue

Long before I contemplated writing this book myself, I wondered why it hadn't already appeared in print. It was easy to envisage a volume on the fauna of Lake Victoria and I couldn't imagine that a nonfiction writer wasn't already writing it somewhere. After several years of waiting, the book still hadn't appeared but my desire to read it persisted. So I began writing it myself. I have chosen to alternate scientific with narrative passages. The biologists who worked with me in Tanzania might wonder why I strove for factual accuracy in the scientific sections but did not feel obliged to do so in the narrative passages. The reason is simply that I did not intend to give a chronological account of anything other than the scientific facts. In some cases the original names of persons and institutes have been changed, in other cases they have been retained.

Many friends and colleagues have contributed in one way or another to the realization of this book. I extend my sincere thanks to all of them, only a few of whom can I mention here by name. Arthur van Leeuwen and Marieke van Oostrom read the manuscript on various occasions with unwaning enthusiasm. Their suggestions and amendments were essential. Maarten Helle corrected numerous editions of the manuscript and provided valuable commentary. I am very grateful to him. I am also indebted to Frans Witte, Cees Barel, Rosemary Lowe-McConnell, Geoffrey Fryer, Kevin Makonda, Jaap de Visser, Monica Soeting, Anne-Mieke Eggenkamp, Kate Simms, Michiel van der Klis, Finette van der Heide, C.J.M. van Herwaarden, Ole Seehausen, Theo

Bakker, Piet Sevenster, Irvin Kornfield, Jacques van Alphen, P. Dullemeijer, K. Bakker, Leslie Kaufman, Richard Ogutu-Owhayo, Jan Wanink, and Kees Goudswaard. Without the support and enthusiasm of the Tanzanian fishermen and the biologists of TAFIRI, the Tanzanian Fisheries and Research Institute at Nyegezi, I would never have stayed so long in East Africa. Naturally, I alone am responsible for the contents of this book.

Glossary

adaptation A change in the form, physiology, or behavior of an organism that occurs in the course of evolution as a result of natural selection and that makes the organism more efficient in disseminating its genes. The changed form, physiological feature, or behavior trait is also called an adaptation.

adaptive radiation The development of a wide variety of species originating from a single ancestral prototype. Well-known old radiations are the marsupials of Australia and the lemurs of Madagascar. Examples of young radiations are the cichlid flocks of the East African lakes, Darwin's finches of the Galápagos Islands, the honeycreepers of Hawaii, and the crustacean-like organisms of Lake Baikal. It is the young radiations that are of particular interest to evolutionary biologists, because phylogenetic relationships and thus often the evolution of form, physiology, and behavior can still frequently be reconstructed.

adenine A purine base; a component of the nucleic acid adenosine, one of the four fundamental constituents of DNA.

allopatric speciation The development of new species after one species has become separated into two or more geographically isolated populations.

allopatry The occurrence of the same species in geographically separate areas (as opposed to sympatry: occurrence in the same area).

amino acid The basic component (for the synthesis) of proteins. Twenty amino acids occur freely in nature, forming the building blocks from which proteins are made.

anemia A shortage of red blood cells; in fact, a shortage of the pigment that gives blood its color (*see* hemoglobin).

automimicry Literally, the imitation of oneself. An example is the egg-dummies on the anal fin of the male furu.

biological species "A population or series of populations within which free gene flow occurs under natural conditions" (definition by E.O. Wilson; *see* chapter 2).

chromosome Carrier of hereditary information. Stainable, thread-like body occurring within nucleus. Composed primarily of nucleic acid.

cichlids Perch-like fish (of the family Cichlidae) to which the species-rich genus *Haplochromis* also belongs.

convergent evolution or **convergence** The development of similar characteristics (anatomical, physiological, behavioral) between two or more unrelated species as a result of natural selection. Best-known examples: whales and fish.

copying behavior Imitation of the choice of another individual instead of choosing independently. Probably plays a role in mate choice.

crypsis The development of a camouflaged exterior.

cytosine A pyrimidine base; a component of the nucleic acid cytidine, one of the four fundamental constituents of DNA.

divergent evolution or **divergence** The branching out of a species in different directions through natural selection. For example, as a result of being exposed to diverse conditions, two populations of the same species develop in different directions.

DNA Deoxyribonucleic acid, a component of the chromosomes containing the hereditary information. Composed of nucleotides.

ecological efficiency The efficiency with which energy is transferred from one link in the food chain to the next.

ecology The study of the relationships between plants and organisms and their physical environment and between the organisms themselves.

ecosystem The plants and organisms occurring in a specific environment (lake, forest, desert), including the physical factors to which they are exposed.

endemic Occurring exclusively within the confines of a clearly defined area. For example, the species of furu in Lake Victoria occur nowhere else except in that lake.

enzymes Substances (usually proteins) produced by organisms that promote or inhibit certain processes. They influence, among other things, various metabolic processes.

ethology The study of the behavior of animals.

evolution The development of life on earth.

evolutionarily stable strategy A strategy that is immune to infiltration by an alternative strategy.

explosive speciation The development of new species at an exceptionally rapid rate. Examples are the furu species in Lake Victoria and the Mbuna species in Lake Malawi.

fitness The ability of an individual to survive and reproduce.

food chain A linear sequencing of prey and their predators. Example: algae, algae-eating crustaceans, crustacean-eating fish, fish-eating fish.

food web The total of food chains in an ecosystem. A graphic representation of a food web gives an idea of the direction of energy flows (and nutrient salts) through an ecosystem.

gene A hereditary unit of a chromosome. Composed of a series of base pairs. In most cases one gene contains the information pertaining to the amino-acid composition of one protein or part of one protein.

genetic code The relationship between the sequence of nucleotides in DNA and the sequence of amino acids in proteins. Important for the process of translation of nucleotide triplets (DNA) into amino acids (proteins).

genetic drift A change in the gene frequencies of a population due to random causes. This phenomenon is particularly significant in small populations, such as those on small islands.

genetic letters The symbols used to indicate the four nucleotides from which DNA is constructed. Each nucleotide has one of the following bases as its primary component: adenine, guanine, cytosine, or thymine.

genome The genetic material of an individual. The total of all the chromosomes and the genes they contain.

genotype All the genes (in each cell) of an individual.

guanine A purine base; a component of the nucleic acid guanosine, one of the four fundamental constituents of DNA.

habitat The environment in which a plant or animal lives.

hemoglobin The protein responsible for the transport of oxygen in the blood.

hemoglobin A or **S** Two forms of hemoglobin. Hemoglobin A is the normal form, hemoglobin S is the deviant form encountered in sickle-cell anemia. Hemoglobin S differs from hemoglobin A in only one amino acid.

heterozygote An individual with two dissimilar genes coding for a specific hereditary characteristic. Put differently, the gene that a child inherits from its mother differs from the one it inherits from its father.

homozygote An individual with two identical genes coding for a specific hereditary characteristic. Put differently, both parents pass on to their child an identical gene of a gene pair.

hybrids Individuals resulting from a cross between individuals of different species.

hybridization The breeding of individuals from different species.

inclusive fitness If "fitness" is the ability of an individual to survive and reproduce, then "inclusive fitness" is the same ability applied to the individual and all its genetic relatives.

isolating mechanism The mechanism preventing fertilization between individuals of different species.

isolation The isolation of species as a result of occupying different ecological niches.

isolation concept A view of biological species based on the notion of the genetic separation of species through isolating mechanisms.

microhabitat The narrowly defined environment occupied by a species. For example, the furu species Haplochromis argens occurs almost exclusively in the uppermost two meters of the water.

mimesis Imitation.

mimic An organism whose outward appearance resembles that of another organism. Examples are edible butterflies that resemble poisonous butterflies, and thereby enjoy a certain degree of protection from predators that avoid the poisonous variety.

mkombozi The Swahili word for "savior." During the mid-1980s, this was the name given to the Nile perch (*Lates* sp.) in Tanzania.

model A poisonous organism with an outward appearance resembling that of an edible organism.

monophyletic group A group of species with a common ancestor. As distinct from a polyphyletic group.

monotypical genus A genus that comprises only a single species.

morphology The study of organic form and structure.

mtDNA Mitochondrial DNA. DNA located outside the nucleus in sausage-shaped cytoplasmic organelles known as mitochondria (important to the production of energy and cell metabolism).

mutation A sudden and random change in the DNA.

natural selection The process whereby forms with a relatively suitable genotype are retained by a population, while those less suited to survival and reproduction disappear. The mechanism of evolution as espoused by Darwin.

niche The role played by a species in the ecosystem.

nucleotides The building blocks of DNA. Four different bases (adenine, guanine, cytosine, and thymine) form the characteristic components of the four different nucleotides from which DNA is built.

nucleus The part of a plant or animal cell containing the chromosomes (composed primarily of DNA).

organelle The part of a cell that can be distinguished by its structure and function.

pelagic Inhabiting the open water, far from the shore (the "littoral" regions).

phenotype The total appearance of an individual organism, as determined by the genotype and the environment.

physiology The study of the processes of living organisms.

phytoplankton Algal plankton (*see* plankton).

plankton Passively drifting microorganisms. Algae form the phytoplankton, microscopically small crustaceans form the zooplankton.

polymerase Enzymes that join together smaller molecules to form larger ones. In the case of DNA polymerase, thousands of nucleotides are joined together into long strands. These are the chromosomal components that carry the genetic information.

polymerase chain reaction A technique by which a minimum quantity of DNA (until recently, too small for investigation) can be artificially amplified into a workable quantity.

polymorphism A variation of forms. The simultaneous occurrence of different forms within a species.

polyphyletic group A group of species descended from different evolutionary lines. As distinct from a monophyletic group.

population A group of members of the same species occurring within a specific area that differ genetically from members of the same species in other areas.

population-genetic species The total of all the organisms that together represent that particular species in nature; the concrete group of organisms within which a free flow of genes occurs. As distinct from the taxonomic species.

primer A small piece of DNA (several dozens of nucleotides long) with a specific nucleotide sequence that allows the polymerase to attach to a strand of DNA at a predetermined position. Without this primer, the polymerase cannot do its work.

proteins Organic materials of high molecular value, occurring readily in nature. Essential to the life processes of plants and animals. Composed of amino acids.

radiation *See* adaptive radiation.

reproductive barrier Ecological, ethological, and/or anatomical barrier that impedes or prevents the hybridization of species.

ribosomes Sites for protein synthesis in the cell. The egg-shaped organelles in the cell in which the amino acids are joined together to form complete proteins.

selection pressure A factor that ensures that some variations of a population survive while others disappear.

sexual selection A distinct form of natural selection that normally occurs as a result of competition between males for females or as a result of choosiness of females in mate choice.

sickle cell A red blood cell having a crescent-like shape. Present in patients suffering from sickle-cell anemia.

sickle-cell anemia A specific form of anemia. In sufferers from sickle-cell anemia, the red blood cells develop a characteristic sickle shape as a result of low blood oxygen levels; consequently, they more readily block small blood vessels and are more vulnerable than red blood cells.

sociobiology The study of the evolution of social behavior.

speciation The appearance of new species.

species *See* biological species.

species flock A group of species descending from a common ancestor and originating within a defined area.

sympatric speciation The development of two new species without one species having first been split into geographically isolated populations.

taxonomic species A species that is specifically defined by a taxonomist for practical reasons, in contrast to the population-genetic species, which defines itself.

taxonomy The science that focuses on the classification of organisms. The goal is to establish a classification that reflects the history of evolution or phylogeny.

thymine Pyrimidine base; a component of the nucleic acid thymidine, one of the four fundamental constituents of DNA.

trophic types Groups of organisms specialized in different areas with respect to food intake.

zooplankton Animal plankton (*see* plankton).

References

Abbreviations Used

Acta Biotheor.: Acta Biotheoretica
Afr. J. Ecol.: African Journal of Ecology
Am. Nat.: The American Naturalist
Am. Zool.: American Zoologist
Anim. Behav.: Animal Behaviour
Annls. Mus. R. Afr. Cent. Sci. Zoo.: Musée Royal de l'Afrique Centrale, Annales Sciences Zoologiques
Behav. Ecol. Sociobiol.: Behavioral Ecology and Sociobiology
Bull. Br. Mus. Nat. Hist. (Zool.): Bulletin of the British Museum of Natural History (Zoology)
E. Afr. For. J.: East African Forestry Journal
E. Afric. J. Agr.: East African Journal of Agriculture
Ecol. Entomol.: Ecological Entomology
Environ. Biol. Fish.: Environmental Biology of Fishes
Int. Rev. Gesamt. Hydrobiol.: Internationale Revue der gesamten Hydrobiologie
J. Exp. Biol.: Journal of Experimental Biology
J. Fish Biol.: Journal of Fish Biology
J. Theor. Biol.: Journal of Theoretical Biology
J. Zool.: Journal of Zoology
Mitt. Int. Verein. Limnol.: Mitteilungen Internationale Vereinigung für theoretische und angewandte Limnologie
Neth. J. Zool.: Netherlands Journal of Zoology
Occas. Pap. Fish. Dept. Uganda: Occasional Papers of the Fisheries Department Uganda
Proc. Nat. Ac. Sci. (USA): Proceedings of the National Academy of Sciences (USA)
Proc. R. Soc. Lond.: Proceedings of the Royal Society of London
SIL Mitteilungen: Societas Internationalis Limnologiae Mitteilungen
Trends Ecol. Evol.: Trends in Ecology and Evolution

Verh. Internat. Verein. Limnol.: Verhandlungen Internationale Vereinigung für theoretische und angewandte Limnologie
Water Qual. Bull.: Water Quality Bulletin
Z. Tierpsychol.: Zeitschrift für Tierpsychologie
Zool. Verh. Leiden: Zoölogische Verhandelingen van het Nationaal Natuurhistorisch Museum te Leiden

1. Greenwood, P. H. (1981). *The Haplochromine Fishes of the East African Lakes.* Kraus Int., Munich.

2. Lévi-Strauss, C. (1962). *La pensée sauvage.* Plon, Paris. Translation: (1966) *The Savage Mind.* University of Chicago Press, Chicago.

3. Barel, C. D. N., Ligtvoet, W., Goldschmidt, T., Witte, F., Goudswaard, P. C. (1991). The haplochromine cichlids in Lake Victoria: an assessment of biological and fisheries interests. In: Keenleyside, M. H. A. (ed.): *Cichlid Fishes, Behaviour, Ecology and Evolution*, Chapman & Hall, London, pp. 258–279.

4. Goldschmidt, T., Witte, F. (1992). Explosive speciation and adaptive radiation of haplochromine cichlids from Lake Victoria: an illustration of the scientific value of a lost species flock. *Mitt. Int. Verein. Limnol.* 23:101–107.

5. Grant, P. R. (1986). *Ecology and Evolution of Darwin's Finches.* Princeton University Press, Princeton.

6. Beadle, L. C. (1981). *The Inland Waters of Tropical Africa.* Longman, London.

7. Keenleyside, M. H. A. (1991). *Cichlid Fishes, Behaviour, Ecology and Evolution.* Chapman & Hall, London.

8. Fryer, G., Iles, T. D. (1972). *The Cichlid Fishes of the Great Lakes of Africa: their Biology and Evolution.* Oliver and Boyd, Edinburgh.

9. Futuyma, D. J. (1986). *Evolutionary Biology.* Sinauer Ass. Inc, Sunderland, Mass.

10. (1977) *Het epos van Heraklios.* Meulenhoff, Amsterdam (Translated from the original Swahili into Dutch by J. Knappert. This excerpt translated from Dutch into English by S. Marx-Macdonald.)

11. Witte, F. (1981). Initial results of the ecological survey of the haplochromine cichlid fishes from the Mwanza Gulf of Lake Victoria (Tanzania): breeding patterns, trophic and species distribution, with recommendations for commercial trawl-fisheries. *Neth. J. Zool.* 31:175–202.

12. Malthus, T. R. (1798). *An assay on the principle of population as it affects the future improvement of society.* John Murray, London.

13. Greenwood, P. H. (1974). The Cichlid Fishes of Lake Victoria, East Africa: the Biology and Evolution of a Species Flock. *Bull. Br. Mus. Nat. Hist. (Zool.)* Supp. 6:1–134.

14. Valéry, P. (1987). *Wat af is, is niet gemaakt.* De Bezige Bij, Amsterdam (This excerpt translated from Dutch into English by S. Marx-Macdonald.)

15. Wilson, E. O. (1988). *Biodiversity*. National Academy Press, Washington, DC.

16. Barel, C. D. N., Oijen, M. J. P. van, Witte, F., Witte-Maas, E. L. M. (1977). An introduction to the taxonomy and morphology of the haplochromine Cichlidae from Lake Victoria. *Neth. J. Zool.* 27:333–389.

17. Barel, C. D. N. (1985). *A Matter of Space. Constructional Morphology of Cichlid Fishes*. Thesis, Rijksuniversiteit Leiden, Leiden, The Netherlands.

18. Hoogerhoud, R. J. C. (1984). A taxonomic reconsideration of the haplochromine genera *Gaurochromis* Greenwood, 1980 and *Labrochromis* Regan, 1920 (Pisces, Cichlidae). *Neth. J. Zool.* 34:539–565.

19. Oijen, M. J. P. van (1991). A revision of the piscivorous haplochromine cichlids of Lake Victoria, Part I. *Zool. Verh. Leiden* 261:1–95.

20. Witte, F., Oijen, M. J. P. van (1990). Taxonomy, ecology and fishery of Lake Victoria haplochromine trophic groups. *Zool. Verh. Leiden* 262:1–67.

21. Yamaoka, K. (1991). Feeding relationships. In: Keenleyside, M. H. A. (ed.): *Cichlid Fishes: Behaviour, Ecology and Evolution*, Chapman and Hall, London, pp. 151–172.

22. Wilhelm, W. (1980). The disputed feeding behaviour of a paedophagous haplochromine cichlid (Pisces) observed and discussed. *Behaviour* 74:310–323.

23. Witte-Maas, E. L. M. (1981). Egg snatching: an observation on the feeding behaviour of *Haplochromis barbarae* Greenwood, 1967 (Pisces, Cichlidae). *Neth. J. Zool.* 31:786–789.

24. Witte, F., Witte-Maas, E. L. M. (1981). Haplochromine cleaner fishes: a taxonomic and eco-morphological description of two new species. Revision of the haplochromine species (Teleostei, Cichlidae) from Lake Victoria. *Neth. J. Zool.* 31:203–231.

25. Nishidi, M. (1991). Lake Tanganyika as an evolutionary reservoir of old lineages of East African cichlid fishes: inferences from allozyme data. *Experientia* 47:974–979.

26. Stiassny, M. L. J. (1991). Phylogenetic intrarelationships of the family Cichlidae: an overview. In: Keenleyside, M. H. A. (ed.): *Cichlid Fishes: Behaviour, Ecology and Evolution*, Chapman and Hall, London, pp. 1–35.

27. Stiassny, M. L. J. (1981). The phyletic status of the family Cichlidae (Pisces, Perciformis): a comparative anatomical investigation. *Neth. J. Zool.* 31:275–314.

28. Kornfield, I. (1991). Genetics. In: Keenleyside, M. H. A. (ed.): *Cichlid Fishes: Behaviour, Ecology and Evolution*, Chapman and Hall, London, pp. 103–128.

29. Meyer, A., Kocher, T., Basasibwaki, P., Wilson, A. C. (1990). Monophyletic origin of Lake Victoria cichlid fishes suggested by mitochondrial DNA sequences. *Nature* 347:550–553.

30. Schellekens, H., ed. (1993). *De DNA-makers. Architecten van het leven.* Natuur & Techniek, Maastricht, The Netherlands.

31. Zolg, W. (1993). De polymerase-kettingreactie. In: Schellekens, H. (ed.): *De DNA-makers. Architecten van het leven*, Natuur & Techniek, Maastricht, The Netherlands, pp. 23–41.

32. Avise, J. Flocks of African fishes (1990). *Nature* 347:512–513.

33. Darwin, C. (1859). *On the Origin of Species by Means of Natural Selection, or the Preservation of Favored Races in the Struggle for Life*. John Murray, London, 6th edition (1894).

34. Williams, G. C. (1992). *Natural Selection: Domains, Levels, and Challenges*. Oxford University Press, Oxford.

35. Endler, J. A. (1986). *Natural Selection in the Wild*. Princeton University Press, Princeton, New Jersey.

36. Fontijn, J. (1994). Biologisch utopisme. Het Darwinisme van Frederik van Eeden. *De negentiende eeuw* 17:33–45 (This excerpt translated from Dutch into English by S. Marx-Macdonald.)

37. Galis, F., Jong, P. W. de (1988). Optimal foraging and ontogeny: food selection by *Haplochromis piceatus*. *Oecologia* 75:175–184.

38. Hori, M. (1993). Frequency-dependent natural selection in the handedness of scale-eating cichlid fish. *Science* 260:216–219.

39. Stryer, L. (1981). *Biochemistry*. W. H. Freeman and Company, San Francisco.

40. Eldredge, N. (1985). *Time Frames. The Rethinking of Darwinian Evolution and the Theory of Punctuated Equilibria*. Simon & Schuster, New York.

41. Rothschild, M. (1972). Colour and poison in insect protection. *New Scientist* 54:318–320.

42. Brower, L. P. (1988). *Mimicry and the Evolutionary Process*. The University of Chicago Press, Chicago.

43. Futuyma, D. J., Slatkin, M. (1983). *Coevolution*. Sinauer Ass. Inc., Sunderland, Mass.

44. Bruggen, C. van (1925). *Hedendaags fetisjisme*. Querido, Amsterdam (This excerpt translated from Dutch into English by S. Marx-Macdonald.)

45. Huheey, J. E. (1988). Mathematical models of mimicry. In: Brower, L. P. (ed.): *Mimicry and the Evolutionary Process*, University of Chicago, Chicago, pp. 22–41.

46. Brower, L. P. (1988). Avian predation on the monarch butterfly and its implications for mimicry theory. In: Brower, L. P. (ed.): *Mimicry and the Evolutionary Process*, University of Chicago, Chicago, pp. 4–7.

47. Wickler, W. (1968). *Mimicry in Plants and Animals*. McGraw-Hill, New York.

48. Guilford, T. (1988). The evolution of conspicuous coloration. In: Brower, L. P. (ed.): *Mimicry and the Evolutionary Process*, University of Chicago, Chicago, pp. 7–21.

49. Brakefield, P. M., Reitsma, N. (1991). Phenotypic plasticity, seasonal climate and the population biology of Bicyclus butterflies (Satyridae) in Malawi. *Ecol. Entomol.* 16:291–303.

50. Mrowka, W. (1987). Egg-stealing in a mouth-brooding cichlid (Pseudocrenilabrus multicolor). *Anim. Behav.* 35:922–923.

51. Wickler, W. (1962). Zur Stammesgeschichte funktionell korrelierter Organ- und Verhaltensmerkmale: Ei-Attrappen und Maulbrüten bei Afrikanischen Cichliden. *Z. Tierpsychol.* 19:129–164.

52. Mrowka, W. (1987). Oral fertilization in a mouthbrooding cichlid fish. *Ethology* 74:293–296.

53. Hert, E. (1989). The function of the egg-spots in an African mouth-brooding cichlid fish. *Anim. Behav.* 37:726–732.

54. McKaye, K. R. (1991). Sexual selection and the evolution of the cichlid fishes of Lake Malawi, Africa. In: Keenleyside, M. H. A. (ed.): *Cichlid Fishes, Behaviour, Ecology and Evolution*, Chapman and Hall, London, pp. 241–257.

55. Goldschmidt, T., Visser, J. de (1990). On the possible role of egg mimics in speciation. *Acta. Biotheor.* 34:125–134.

56. Cohen, A., Johnston, M. (1987). Speciation in brooding and poorly dispersing lacustrine organisms. *Palaios* 2:426–435.

57. Paterson, H. E. H. (1985). The recognition concept of species. In: Vrba, E. S. (ed.): *Species and Speciation*, Transvaal Museum, Pretoria, RSA, pp. 21–29.

58. Mayr, E. (1963). *Animal Species and Evolution.* Harvard University Press, Cambridge, Mass.

59. Crapon de Caprona, M. D., Fritzsch, B. (1984). Interspecific fertile hybrids of haplochromine Cichlidae (Teleostei) and their possible importance for speciation. *Neth. J. Zool.* 34:503–538.

60. Goodall, J. (1986). *The Chimpanzees of Gombe.* Harvard University Press, Cambridge, Mass.

61. Grant, V. (1985). *The Evolutionary Process. A Critical Review of Evolutionary Theory.* Columbia University Press, Columbia.

62. Hoogerhoud, R. J. C., Witte, F., Barel, C. D. N. (1983). The ecological differentiation of two closely resembling *Haplochromis* species from Lake Victoria (*H. iris* and *H. hiatus*; Pisces, Cichlidae). *Neth. J. Zool.* 33:283–305.

63. Ribbink, A. J. (1994). Alternative perspectives on some controversial aspects of cichlid fish speciation. In: Martens, K., Coulter, G., Boodesis, B. (eds.): *Conference Volume on Speciation in Ancient Lakes.*

64. White, M. J. D. (1978). *Modes of Speciation.* W. H. Freeman and Company, San Francisco.

65. Schliewen, U. K., Tautz, D., Paabo, S. (1994). Sympatric speciation suggested by monophyly of crater lake cichlids. *Nature* 368:629–632.

66. Balon, E. K. (1985). *Early Life Histories of Fishes. New Developmental, Ecological and Evolutionary Perspectives.* The Hague.

67. Owen, R. B., Crossley, R., Johnson, T. C., Tweddle, D., Kornfield, I., Davison, S., Eccles, D. H., Engstrom, D. E. (1990). Major low levels of Lake

Malawi and their implications for speciation rates in cichlid fishes. *Proc. R. Soc. Lond.* 240:519–553.

68. Hillenius, D. (1976). *De vreemde eilandbewoner*. Arbeiderspers, Amsterdam.

69. Goldschmidt, T., Witte, F. (1990). Reproductive strategies of zooplanktivorous haplochromine cichlids (Pisces) from Lake Victoria before the Nile perch boom. *Oikos* 58:356–368.

70. Liem, K. F., Osse, J. W. M. (1975). Biological versatility, evolution and food resource exploitation in African cichlid fishes. *Am. Zool.* 15:427–454.

71. Dullemeijer, P. (1974). *Concepts and Approaches in Animal Morphology*. Van Gorcum, Assen, The Netherlands.

72. Liem, K. F., Kaufman, L. S. (1984). Intraspecific macroevolution: functional biology of the polymorphic cichlid species *Cichlasoma minckleyi*. In: Echelle, A. A., Kornfield, I. (eds.): *Evolution of Fish Species Flocks*, University of Maine at Orono Press, Orono, Maine, pp. 203–215.

73. Darwin, C. (1871). *The Descent of Man and Selection in Relation to Sex*. John Murray, London.

74. Fisher, R. A. (1930). *The Genetical Theory of Natural Selection*. Dover, New York.

75. Dominey, W. J. (1984). Effects of sexual selection and life history on speciation: species flocks in African cichlids and Hawaiian *Drosophila*. In: Echelle, A. A., Kornfield, I. (eds.): *Evolution of Fish Species Flocks*, University of Maine at Orono Press, Orono, Maine, pp. 231–249.

76. Endler, J. A. (1988). Sexual selection and predation risk in guppies. *Nature* 332:593–594.

77. Borgerhoff-Mulder, M. (1988). Kipsigis bridewealth payments. In: Betzig, L., Borgerhoff-Mulder, M., Turke, P. (eds.): *Human Reproductive Behaviour. A Darwinian Perspective*, Cambridge University Press, Cambridge, pp. 65–82.

78. Dawkins, R. (1976). *The Selfish Gene*. Oxford University Press, Oxford.

79. Hamilton, W. D. (1964). The genetical theory of social behaviour. *J. Theor. Biol.* 7:1–52.

80. Hart, M. 't (1978). *De stekelbaars*. Het Spectrum, Utrecht, The Netherlands.

81. Sevenster, P. (1961). A causal analysis of a displacement activity (fanning in *Gasterosteus aculeatus* L.). *Behaviour* (Supp.), 9:1–170.

82. Goldschmidt, T., Bakker, Th. C. M. (1990). Determinants of reproductive success of male sticklebacks in the field and in the laboratory. *Neth. J. Zool.* 40:664–687.

83. Goldschmidt, T. (1991). Egg mimics in haplochromine cichlids (Pisces, Perciformes) from Lake Victoria. *Ethology* 88:177–190.

84. Malte Anderson, S. S. (1994). *Sexual Selection*. Princeton University Press, Princeton.

85. Bakker, Th. C. M. (1993). *The Evolution of Male Ornamentation through Female Choice in Three-Spined Sticklebacks, Gasterosteus aculeatus L.*, Habilitationsschrift, University of Bern, Switzerland.

86. Lande, R. (1982). Rapid origin of sexual isolation and character divergence in a cline. *Evolution* 36:213–223.

87. Wilkinson, G. S., Reillo, P. R. (1994). Female choice response to artificial selection on an exaggerated male trait in a stalk-eyed fly. *Proc. R. Soc. Lond. B 255:1–6.*

88. Céline, L. F. (1932). *Voyage au bout de la nuit.* Hatier, Paris. Translation: (1934) *Journey to the End of the Night.* New Directions, New York.

89. Zahavi, A. (1975). Mate selection, a selection for a handicap. *J. Theor. Biol.* 53:205–214.

90. Krebs, J. R., Davies, N. B. (1991). *Behavioural Ecology. An Evolutionary Approach.* Blackwell, Oxford.

91. Hamilton, W. D., Zuk, M. (1982). Heritable true fitness and bright birds: a role for parasites? *Science* 218:384–387.

92. Cronly-Dillon, J., Sharma, S. C. (1968). Effect of season and sex on the phototopic sensitivity of the three-spined stickleback. *J. Exp. Biol.* 49:679–687.

93. Milinski, M., Bakker, Th. C. M. (1990). Female sticklebacks use male coloration in mate choice and hence avoid parasitized males. *Nature* 344:330–333.

94. Pruett-Jones, S. G. (1992). Independent versus nonindependent mate choice: do females copy each other? *Am. Nat.* 140:1000–1009.

95. Milinski, M., Bakker, Th. C. M. (1992). Costs influence sequential mate choice in sticklebacks. *Proc. R. Soc. Lond.* 250:229–233.

96. Dugatkin, L. A., Godin, J. G. J. (1992). Reversal of female mate choice by copying in the guppy (*Poecilia reticulata*). *Proc. R. Soc. Lond.* 249:179–184.

97. Gibson, R. M., Höglund, J. (1992). Copying and sexual selection. *Trends Ecol. Evol.* 7:229–231.

98. Ridley, M., Rechten, C. (1981). Female sticklebacks prefer to spawn with males whose nests contain eggs. *Behaviour* 76:152–161.

99. Jamieson, I. G., Colgan, P. W. (1989). Eggs in the nests of males and their effects on mate choice of female sticklebacks. *Anim. Behav.* 38:859–865.

100. Tinbergen, N. (1951). *The Study of Instinct.* Oxford University Press, Oxford.

101. Assem, J. van den (1967). Territory in the three-spined stickleback, *Gasterosteus aculeatus* L. An experimental study in intra-specific competition. *Behaviour* (Supp.) 16:1–194.

102. Goldschmidt, T., Bakker, Th. C. M., Feuth-de Bruijn, E. (1993). Selective copying in mate choice of female sticklebacks. *Anim. Behav.* 45:541–547.

103. Goldschmidt, T., Foster, S. A., Sevenster, P. (1992). Inter-nest distance and sneaking in the three-spined stickleback. *Anim. Behav.* 44:793–795.

104. Emlen, S. T., Oring, L. W. (1977). Ecology, sexual selection, and the evolution of mating systems. *Science* 197:215–223.

105. Crook, J. H. (1964). The evolution of social organization and visual communication in the weaver birds (Ploceinae). *Behaviour* (Supp.) 10:1–178.

106. Rhijn van, J. G., Westerterp-Plantenga, M. S. (1989). *Ethologie: veroorzaking, ontwikkeling, functie en evolutie van gedrag.* Wolters-Noordhoff, Groningen, The Netherlands.

107. Davies, N. B. (1991). Mating systems. In: Krebs, J. R., Davies, N. B. (eds.): *Behavioural Ecology. An Evolutionary Approach*, Blackwell, Oxford, pp. 263–294.

108. Michaux, H. (1933). *Un Barbare en Asie*. Gallimard, Paris (This excerpt translated from French into English by S. Marx-Macdonald.)

109. Crook, J. H., Crook, S. J. (1988). Tibetan polyandry: problems of adaptation and fitness. In: Betzig, L., Borgerhoff-Mulder, M., Turke, P. (eds.): *Human Reproductive Behaviour, a Darwinian Perspective*, Cambridge University Press, New York, pp. 97–114.

110. Barlow, G. W. (1991). Mating systems among cichlid fishes. In: Keenleyside, M. H. A. (ed.): *Cichlid Fishes, Behaviour, Ecology and Evolution*, Chapman & Hall, London, pp. 173–191.

111. Taborsky, M. (1984). Broodcare helpers in the cichlid fish *Lamprologus brichardi*: their costs and benefits. *Anim. Behav.* 32:1236–1252.

112. Baerends, G. P. (1986). The functional organisation of the reproductive behaviour in cichlid fish. *Annls. Mus. R. Afr. Cent. Sci. Zool.* 251:3–5.

113. Keenleyside, M. H. A. (1991). Parental care. In: Keenleyside, M. H. A. (ed.): *Cichlid Fishes: Behaviour, Ecology and Evolution*, Chapman and Hall, London, pp. 191–208.

114. Sato, T. (1994). A brood parasitic catfish of mouthbrooding cichlid fishes in Lake Tanganyika. *Nature* 323:58–59.

115. Sato, T. (1994). Active accumulation of spawning substrate: a determinant of extreme polygyny in a shell-brooding cichlid fish. *Anim. Behav.* 48:669–678.

116. Le Bouef, B. J. (1974). Male-male competition and reproductive succes in elephant seals. *Am. Zool.* 14:163–176.

117. Collias, N. E., Collias, E. C. (1984). *Nest Building and Bird Behaviour.* Princeton University Press, Princeton.

118. Russel, C., Russel, W. M. S. (1990). Cultural evolution of behaviour. *Neth. J. Zool.* 40:745–762.

119. Barash, D. (1987). *The Hare and the Tortoise: Culture, Biology, and Human Nature.* Viking, New York.

120. Odijk, T., Oosterhoff, B. (1992). *Sociaal-economische en culturele veranderingen in Igombe, een vissergemeenschap in Tanzania tengevolge van de introductie van nijlbaars in het Victoriameer.* Thesis, Katholieke Universiteit, Nijmegen, The Netherlands.

121. Lack, D. (1947). *Darwin's Finches*. Cambridge University Press, Cambridge.

122. Kruuk, H. (1967). Competition for food between vultures in East Africa. *Ardea* 3–4:170–193.

123. Wilson, E. O. (1992). *The Diversity of Life*. The Belknap Press, Cambridge, Mass.

124. Ehrlich, P. R. (1986). *The Machinery of Nature*. Simon & Schuster, New York.

125. MacArthur, R. H. (1972). *Geographical Ecology*. Harper and Row, New York.

126. Goldschmidt, T., Witte, F., Visser, J. de (1990). Ecological segregation in zooplanktivorous haplochromine species (Pisces, Cichlidae) from Lake Victoria. *Oikos* 58:343–355.

127. Witte, F. (1984). Ecological differentiation in Lake Victoria haplochromines: comparison of cichlid species flocks in African lakes. In: Echelle, A. A., Kornfield, I. (eds.): *Evolution of Fish Species Flocks*, University of Maine at Orono Press, Orono, Maine, pp. 155–167.

128. Meer, H. J., van der, Anker, G. Ch. (1984). Retinal resolving power and sensitivity of the photopic system in seven haplochromine species (Cichlidae, Teleostei). *Neth. J. Zool.* 34:137–209.

129. Schoener, T. W. (1989). The ecological niche. In: Cherett, J. M. (ed.): *Ecological Concepts*, Blackwell Scientific Publications, Oxford, pp. 79–113.

130. Strong, D. R., Simberloff, D., Abele, L. G., Thistle, A. B. (1984). *Ecological Communities: Conceptual Issues and the Evidence*. Princeton University Press, Princeton.

131. Chesson, P. L., Huntly, N. (1989). Short-term instabilities and long-term community dynamics. *Trends Ecol. Evol.* 4:293–298.

132. Anderson, A. M. (1961). Further observation concerning the proposed introduction of Nile perch into Lake Victoria. *E. Afr. For. J.* 26:195–201.

133. Fryer, G. (1960). Concerning the proposed introduction of Nile perch into Lake Victoria. *E. Afric. J. Agr.* 25:267–270.

134. Lowe-McConnell, R. H. (1993). Fish faunas of the African Great Lakes: origins, diversity, and vulnerability. *Conservation Biology* 7:634–643.

135. Kinloch, B. (1972). *The Shamba Raiders, Memories of a Game Warden*. Collins and Harvill Press, London.

136. Ogutu-Ohwayo, R. (1990). The decline of the native fishes of Lake Victoria and Kyoga (East Africa) and the impact of introduced species, especially the Nile perch, *Lates niloticus*, and the Nile tilapia, *Oreochromis niloticus*. *Environ. Biol. Fishes* 27:81–96.

137. Ogutu-Ohwayo, R. (1990). The reduction in fish species diversity in Lakes Victoria and Kyoga (East Africa) following human exploitation and introduction of non-native fishes. *J. Fish. Biol.* 37:207–208.

138. Stoneman, J., Rogers, J. F. (1970). Increase in fish production achieved by stocking exotic species (Lake Kyoga, Uganda). *Occas. Pap. Fish. Dept. Uganda* 3:16–19.

139. Raup, D. M. (1993). *Extinction, Bad Genes or Bad Luck?* Oxford University Press, Oxford.

140. Prins, R. A., Emden, H. M. van (1989). *Het verdwijnen van soorten.* KNAW, Amsterdam.

141. Foose, T. J. (1994). Riders of the last ark: the role of captive breeding in conservation strategies. In: Kaufman, L., Mallory, K. (eds.): *The last extinction*, The MIT Press, Cambridge, pp. 149–178.

142. Smit, J. (1989). Een catastrofale milieucrisis 66. 5 miljoen jaar geleden. In: Prins, R. A., Emden, H. M. van (eds.): *Het verdwijnen van soorten*, KNAW, Amsterdam, pp. 41–59.

143. Raup, D. M., Sepkoski, J. J. (1984). Periodicity of extinctions in the geologic past. *Proc. Nat. Ac. Sci. (USA)* 81:801–805.

144. Raup, D. M., Sepkoski, J. J. (1986). Periodic extinction of families and genera. *Science* 231:833–836.

145. Patterson, C., Smith, A. B. (1989). Periodicity in extinction: the role of systematics. *Ecology* 70:802–811.

146. Hughes, N. F. (1983). *A Study of the Nile Perch, an Introduced Predator in the Kavirondo Gulf of Lake Victoria.* Oxford University Press, Oxford.

147. Witte, F., Goldschmidt, T., Wanink, J., Oijen, M. J. P. van, Goudswaard, P. C., Witte-Maas, E. L. M., Boutton, N. (1992). The destruction of an endemic species flock: quantitative data on the decline of the haplochromine cichlids of Lake Victoria. *Environ. Biol. Fishes* 34:1–28.

148. Brönmark, C., Miner, J. G. (1992). Predator-induced phenotypical change in body morphology in Crucian carp. *Science* 258:1348–1350.

149. Kaufman, L. (1993). Why the ark is sinking. In: Kaufman, L., Mallory, K. (eds.): *The Last Extinction*, The MIT Press, Cambridge, pp. 1–41.

150. Reyer, H. U. (1980). Flexible helper structure as an ecological adaptation in the pied kingfisher (*Ceryle rudis rudis* L.). *Behav. Ecol. Sociobiol.* 6:219–227.

151. Witte, F., Goldschmidt, T., Wanink, J. (1995). Dynamics of the Haplochromine cichlid fauna and other ecological changes in the Mwanza Gulf of Lake Victoria. In: Pitcher, T. J., Hart, P. J. B. (eds.): *Impact of Species Changes on African Lakes*, Chapman & Hall, London, pp. 83–110.

152. Barel, C. D. N., Dorit, P., Fryer, G., Hughes, N., Jackson, P. B. N., Kawanabe, H., Lowe-McConnell, R. H., Nagoshi, M., Trewavas, E., Witte, F., Yamaoka, K. (1985). Destruction of fisheries in Africa's Lakes. *Nature* 315:19–20.

153. Meyers, D. G., Stricker, J. R. (1984). *Trophic Interactions within Aquatic Ecosystems*. Westview Press, Colorado.

154. Mkumbo, O. C., Ligtvoet, W. (1992). Changes in the diet of Nile perch, Lates niloticus (L.) in the Mwanza Gulf, Lake Victoria. *Hydrobiologia* 232:79–83.

155. Goldschmidt, T., Witte, F., Wanink, J. (1993). Cascading effects of the introduced Nile perch on the detritivorous/phytoplanktivorous species in the sublittoral areas of Lake Victoria. *Conservation Biology* 7:686–700.

156. Ogari, J., Dadzie, S. (1988). The food of Nile perch, Lates niloticus (L.), after the disappearance of the haplochromine cichlids in Nyanza Gulf of Lake Victoria (Kenya). *J. Fish. Biol.* 32:571–577.

157. Wanink, J. (1991). Survival in a perturbed environment: the effects of Nile perch introduction on the zooplanktivorous fish community of Lake Victoria. In: Ravera, O. (ed.): *Terrestrial and Aquatic Ecosysytems: Perturbation and Recovery*, Ellis Horwood, New York, pp. 269–275.

158. Ligtvoet, W., Witte, F. (1993). Perturbation by predator introduction: effects on trophic structure and fisheries yield in Lake Victoria (East Africa). In: Ravera, O. (ed.): *Terrestrial and Aquatic Ecosystems: Perturbation and Recovery*, Ellis Horwood Ltd., Chichester, pp. 263–268.

159. Carpenter, S. R., Kitchell, J. F., Hodgson, J. R. (1985). Cascading trophic interactions and lake productivity. *Bioscience* 35:634–639.

160. Ochumba, P. B. O., Kibaara, D. I. (1989). Observations on the blue-green algal blooms in the open waters of Lake Victoria, Kenya. *Afr. J. Ecol.* 27:23–34.

161. Kaufman, L. S. (1992). Catastrophic change in a species-rich freshwater ecosystem: the lessons of Lake Victoria. *Bioscience* 42:846–858.

162. Talling, J. F. (1966). The annual cycle of stratification and phytoplankton growth in Lake Victoria, East Africa. *Int. Rev. Gesamt. Hydrobiol.* 51:545–621.

163. Kaufman, L., Ochumba, P. B. O. (1993). Evolutionary and conservation biology of cichlid fishes as revealed by faunal remnants in northern Lake Victoria. *Conservation Biology* 7:719–729.

164. Cohen, A. (1992). Criteria for developing underwater reserves in Lake Tanganyika. SIL. *Mitteilungen* 23:109–116.

165. Andrews, C. (1992). The role of zoos and aquaria in the conservation of the fishes from Africa's Great Lakes. *Mitt. Int. Verein. Limnol.* 23:117–120.

166. Hecky, R. E. (1993). The eutrophication of Lake Victoria. *Verh. Internat. Verein. Limnol.* 25:39–48.

167. Twongo, T. (1993). The spread of water hyacinth on Lakes Victoria and Kyoga and some implications for aquatic biodiversity and fisheries. *People, Fisheries, Biodiversity and the Future of Lake Victoria* 93–3:42.

168. Ochumba, P. B. O. (1987). Periodic massive fish kills in the Kenyan part of Lake Victoria. *Water Qual. Bull.* 12:119–122.

Index